Ionosphere - New Perspectives

Edited by Yann-Henri H. Chemin

Published in London, United Kingdom

Ionosphere - New Perspectives
http://dx.doi.org/10.5772/intechopen.1001520
Edited by Yann-Henri H. Chemin

Contributors
Omar A. Fadhil, Vladimir Pashintsev, Mark Peskov, Dmitry Mikhailov, Mikhail Senokosov, Dmitry Solomonov, Myrto Tzamali, Spiros Pagiatakis, Nouf Abd Elmunim, Chengli She, Adriano Camps, Carlos Molina, Guillermo González-Casado, José Miguel Juan, Joël Lemorton, Vincent Fabbro, Aymeric Mainvis, José Barbosa, Raúl Orús-Pérez, Sergey Matsievsky, Wayne Scales, Augustine Yellu

First published in London, United Kingdom, 2023 by IntechOpen
IntechOpen is the global imprint of INTECHOPEN LIMITED, registered in England and Wales, registration number: 11086078, 5 Princes Gate Court, London, SW7 2QJ, United Kingdom

British Library Cataloguing-in-Publication Data
A catalogue record for this book is available from the British Library

Additional hard and PDF copies can be obtained from orders@intechopen.com

Ionosphere - New Perspectives
Edited by Yann-Henri H. Chemin
p. cm.
Print ISBN 978-1-83769-538-6
Online ISBN 978-1-83769-537-9
eBook (PDF) ISBN 978-1-83769-539-3

Meet the editor

Dr. Yann-Henri Chemin has a BSc in Planetary Sciences with Astronomy from Birkbeck, University of London, England, an MSc in Land and Water Resources Management from Cranfield University, United Kingdom, and a DTSc in Remote Sensing and GIS applications and an eMBA in Security, Defense and Space Industries from IAE Nice, France. He has worked on modeling thermodynamic fluxes from remotely sensed imagery throughout the Earth's atmosphere and on the hyperspectral mapping of the Apollo 12 landing site. He has edited several books on Earth remote sensing, lunar science, astronomy, cryovolcanism, solar wind, and genetic algorithms.

Contents

Preface

The ionosphere is a layer of the Earth's atmosphere that extends from about 50 km to 1000 km above the Earth's surface. It is called the ionosphere because it is ionized by solar radiation, which creates ions and free electrons in the upper atmosphere. These ions and electrons reflect radio waves back to the Earth's surface, allowing long-distance radio communication and absorption of harmful solar radiation, thus protecting life on Earth.

One of the primary effects of solar wind on the ionosphere is the enhancement of the ionospheric plasma density. When solar wind interacts with the Earth's magnetosphere, it compresses the magnetosphere and creates a region of increased ionization in the ionosphere. This region is known as the plasmasphere and can extend up to six Earth radii from the Earth's surface. The plasmasphere is composed of cold, dense plasma that can absorb and reflect radio waves, leading to communication blackouts and loss of navigational satellite signals.

Solar wind can also cause the ionosphere to expand, particularly during periods of high solar activity, such as solar flares and coronal mass ejections. The enhanced ionization in the ionosphere can increase the drag on low-Earth orbiting satellites and cause them to decay faster. The expansion of the ionosphere can also affect radio wave propagation, leading to signal delay and distortion. Therefore, the ionosphere must be continuously monitored to predict and mitigate the effects of ionospheric storms.

Ionospheric weather

There are different types of ionospheric weather conditions, each of which has its own characteristics and impact on satellite communications. The most common ones include:

1. *Space radiation*: light from the solar and lunar sources that is absorbed by the ionosphere, modifying its transparency.

2. *Space radiation noise*: a type of noise that is emitted by the ionosphere and is caused by a variety of sources, including sunlight, radiation from the ground, and noise in power lines.

3. *Space vicinity weather conditions*: conditions that occur in the ionosphere surrounding the orbiter or space vehicle. Charged particles from the satellite collide with the space environment, which can result in ionization or modification of the platform's signals.

4. *Drifter conditions*: these are due to the movement of sand particles and dust in the ionosphere. Drifters are used in satellite navigation systems to control the location of a satellite over long distances.

Ionospheric conductivity monitoring assesses the state of the ionosphere and improves the accuracy of satellite communications. It is used in ground-based communications systems to allow for remote monitoring of communications.

Composing layers of the ionosphere

The D-layer is the lowest layer of the ionosphere, located between 60 km and 90 km above the Earth's surface. It is formed by the absorption of high-frequency radio waves by free electrons in the ionosphere. It is also known as the daytime absorptive layer because it is most effective during daylight hours. It can have a significant impact on high-frequency radio communication systems, as it absorbs radio waves with frequencies less than 10 MHz, making long-distance communication challenging.

The E-layer is the second lowest layer of the ionosphere, located between 90 km and 150 km above the Earth's surface. The E-layer is formed by the ionization of neutral atoms in the Earth's atmosphere by solar radiation. The E-layer is also known as the Heaviside layer after Oliver Heaviside, who first proposed the existence of a reflecting layer in the atmosphere. The E-layer has a significant impact on medium-frequency radio communication systems, as it reflects radio waves in the 3 to 30 MHz range. The E-layer is strongest during the daytime and is affected by the solar cycle, with its maximum height and electron density occurring during periods of high solar activity.

The F-layer is the highest layer of the ionosphere, located between 150 km and 1000 km above the Earth's surface. The F-layer is composed of two sub-layers, F1 and F2, which are separated by a region of minimum electron density known as the F-region trough. The F1-layer is located between 150 km and 200 km above the Earth's surface and is formed by the ionization of neutral atoms in the Earth's atmosphere by solar radiation. The F1-layer has a significant impact on very high-frequency radio communication systems, as it reflects radio waves in the 30 to 50 MHz range. The F1-layer is strongest during the daytime and is affected by the solar cycle, with its maximum height and electron density occurring during periods of high solar activity.

The F2-layer is located between 200 km and 1000 km above the Earth's surface and is formed by the ionization of neutral atoms in the Earth's atmosphere by solar radiation. The F2-layer is the strongest and most important layer of the ionosphere, as it has the highest electron density and reflects radio waves in the very high-frequency (VHF) and ultra-high frequency (UHF) range. The F2-layer has a significant impact on communication and navigation systems, such as satellite communication and GPS systems. The F2-layer is also affected by the solar cycle, with its maximum height and electron density occurring during periods of high solar activity.

This book provides a comprehensive overview of the ionosphere in two sections. The first section examines the influence and impact of transient or orbiting humanmade objects into the ionosphere. The second section discusses the monitoring and modeling of the temporal evolution of the ionosphere, leading to a better understanding and forecasting of its dynamic.

Yann-Henri Chemin
European Comission, JRC
Ispra, Italy

Section 1

Ionospheric Interactions with Man-Made Orbiters

Chapter 1

Entry Dynamics of Space Shuttle in the Earth Atmosphere

Omar A. Fadhil

Abstract

There are two entry methods that control the design of vehicles to return to the atmosphere from outer space: ballistic entry and lifting entry. In this chapter, we will study the process of entering the space shuttle into the Earth's atmosphere. In the case of ballistic entry without lift at sufficiently large angle of inclination, both the gravity force and centrifugal force are neglected. The term ballistic entry is applied to the relatively steep atmosphere entry of non-lifting bodies, which involves linear paths through the atmosphere during the major deceleration, with assumptions: (1) A constant path angle. (2) A constant drag coefficient. (3) Gravitational force small compared to drag force. Program was designed by the author to calculate variation of velocity with altitude for various values of drag-weight parameter, variation of deceleration with altitude for various values of drag-weight parameter, variation of velocity with altitude for various values of initial angle of inclination, variation of deceleration with altitude for various values of initial angle of inclination, variation of velocity with altitude for various values of initial velocity and variation of deceleration with altitude for various values of initial velocity.

Keywords: atomspheric drag, entry flight path angle, space shuttle entry, inclination, deceleration

1. Introduction

Atmospheric entry is the movement of an object coming from space into the atmosphere of a particular planet. The process of entry is divided into two types, namely regular entry, as in spacecraft, and irregular entry, as in exploding meteorites, etc..

Objects that enter the atmosphere are subjected to a drag process due to the particles that make up the atmosphere, which causes a loss of mass for the entering body or may cause an explosion to some objects that cannot withstand high pressure. This depends on the entry speed and entry angle, as well as the density of the atmosphere [1].

There are two types of re-entry process from space to earth atmosphere:

a. Ballistic entry: This type of entry depends on the parameter ballistic coefficient β and can be calculated from equation below [2]:

$$\beta = \frac{W}{C_D A} \tag{1}$$

where W: the vehicle weight; C_D: the drag coefficient; A: the reference area used in the definition of the drag coefficient.

If the value of β is low, this means that heating and deceleration are less than for a high β value. Early Inter-Continental Ballistic Missiles (ICBM) is one of the types of ballistic missiles that need to leave the atmosphere, then go to the desired target, and then re-enter the atmosphere for a second.

How is the launch process? Upon takeoff, the ICBM enters the launch phase. During that stage, the auxiliary missiles that push the ICBM into the atmosphere are carried upwards, for a period of 2–5 min, until it reaches space. He adds that the ICBM can have up to three stages of auxiliary missiles. Each stage is disposed of after it burns; in other words, after the first stage stops burning, the auxiliary rocket No. 2 takes its turn, and so on. Moreover, these auxiliary missiles can have liquid or solid fuel. Liquid propellants continue to burn longer in the launch phase than solid propellant auxiliary rockets. In contrast, solid fuels "save their energy in a shorter time and burn faster".

b. Lifting entry

A lifting entry show in **Figure 1** is one in which the primary force being generated is perpendicular to the flight path, that is, a "lift" force. Although drag is present throughout the entry, the resulting flight path can be adjusted continuously to change both vertical motion and flight direction, while the velocity is slowing. The gliding flight of a sailplane is an example of "lifting" entry without high velocities and heating. The primary design parameter for lifting entry is the lift to drag ratio, or L/DA lifting entry is one in which the primary force being generated is perpendicular to the flight path, that is, a "lift" force. Although drag is present throughout the entry, the resulting flight path can be adjusted continuously to change both vertical motion and flight direction while the velocity is slowing. The gliding flight of a sailplane is an example of "lifting" entry without high velocities and heating. The primary design parameter for lifting entry is the lift to drag ratio, or L/D

$$\mathrm{L/D} = \mathrm{Lift/Drag} = C_L/C_D$$

Low L/D values result in low mobility, moderate heating levels, and moderate g loads. High L/D values result in very low g loads but extremely long-lasting entry with continual heating. As an illustration, consider the space shuttle's re-entry, which took around 25 min overall and had an L/D value of about unity [3]. Although a lifting entrance's peak temperatures are lower than a ballistic entry's peak temperatures, the overall heat load that must be absorbed throughout the entry is larger. As the L/D increases, lateral mobility during entry (often referred to as "cross-range capability") grows significantly.

2. Dynamics and vehicle description

The 3-DOF point mass dynamics of the reentry vehicle over a spherical, rotating Earth, is described as follows:

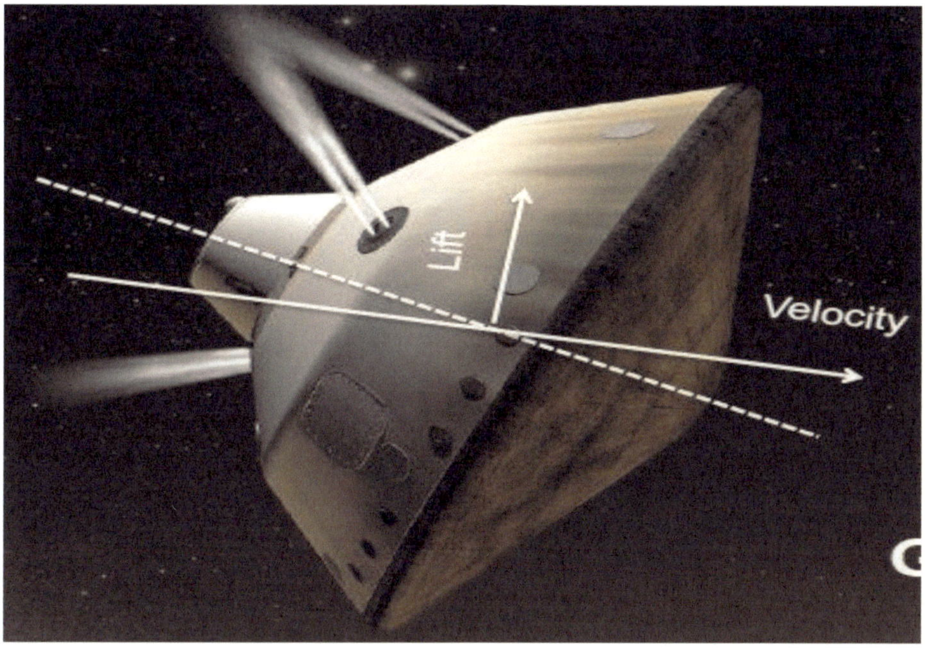

Figure 1.
Lifting entry.

$$\dot{r} = V \sin\gamma, \tag{2}$$

$$\dot{\theta} = \frac{V \cos\gamma \sin\psi}{r\cos\varphi} \tag{3}$$

$$\dot{\varphi} = \frac{V \cos\gamma \cos\psi}{r} \tag{4}$$

$$\dot{V} = -\frac{D}{m} - g \sin\gamma + \omega^2 r \cos\varphi(\sin\gamma\cos\varphi - \cos\gamma\sin\varphi\cos\psi) \tag{5}$$

$$\dot{\gamma} = \frac{1}{V}\left(\frac{L \cos\delta}{m} + \left(\frac{V^2}{r} - g\right)\cos\gamma + 2\omega V\cos\varphi\sin\psi \right. \tag{6}$$

$$\left. + \omega^2 r\cos\varphi(\cos\gamma\cos\varphi + \sin\gamma\sin\varphi\cos\psi)\right),$$

$$\dot{\psi} = \frac{1}{V}\left(\frac{L \sin\delta}{m} + \frac{V^2}{r}\cos\gamma\sin\psi\tan\varphi - 2\omega V(\cos\varphi\tan\gamma\cos\psi - \sin\varphi) + \left(\frac{\omega^2 r}{\cos\gamma}\sin\varphi\cos\varphi\sin\psi\right)\right), \tag{7}$$

where r is the earth radius (the distance from center of earth to vehicle), h is latitude, the earth radius is 6378 m, and θ and φ are the longitude and latitude, respectively.

V is earth relative velocity, γ is the flight path angle of the earth relative velocity, ψ is the azimuth angle of the earth relative velocity, and m is the mass of vehicle [4].

3. System implications

The re-entry system designer can consider three main variables:

1. Ballistic coefficient β.

2. L/D for lift to drag.

3. Entry flight path angle.

Which regulate the re-entry system's flight performance. We will now provide an illustration of how these criteria can be applied to the design of both ballistic and maneuverable warhead delivery systems. Slender sphere–cone geometries with several warheads on a single delivery bus are frequently used in modern ballistic trajectory nuclear weapon system delivery vehicles. The re-entry vehicle base diameter and vehicle length are chosen by the designer for a particular warhead and accompanying arming system, thus determining the cone half-angle [5]. **Figure 2** shows the impacts of nose bluntness ratio on sphere–cone drag coefficient CD for a family of sphere-cone half-angles on classical Newtonian theory, and the nose bluntness ratio is then chosen based on drag and heat transfer concerns [6].

The combined weight of the warhead, arming device, and re-entry vehicle serves as the vehicle's fixed weight. A designer can now evaluate the ballistic trajectory performance of the design in relation to mission criteria, such as deceleration g loads, range, and flight time, since the ballistic coefficient β has been thoroughly determined. These mission requirements actually determine the entry flight path angle.

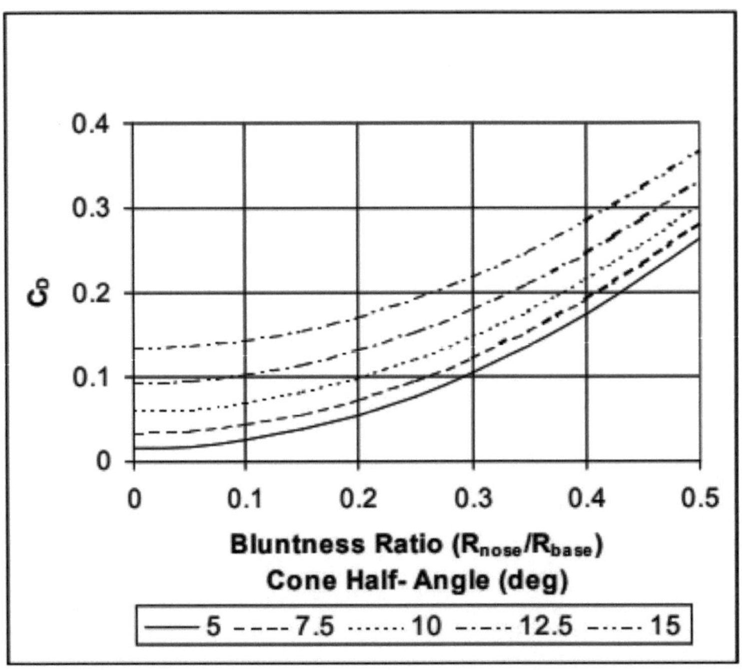

Figure 2.
Classical Newtonian drag coefficient for a sphere-cone.

Before all of the mission requirements are fully met, this process may need to be iterated on numerous times [7].

4. Results and discussion

4.1 Introduction

Trajectory analysis of re-entry vehicles has been modeled in two.

The dynamic equations of motion in 2DOF were solved numerically by adopting the fourth order Runge-Kutta method. Different values of lift-drag ratio were taken (positive, zero, and negative values). And different angles of inclination were taken (small, median, and large values), as well as different weight-area ratios and density–altitude. Further, the dynamic equations of motion in 2DOF were solved analytically. There are four cases for analytical solution.

4.2 Ballistic entry (L/D = 0) at large angles of inclination

In the case of ballistic entry without lift at sufficiently large angle of inclination, both the gravity force and centrifugal force are neglected.

The term ballistic entry is applied to the relatively steep atmosphere entry of non-lifting bodies, which involves linear paths through the atmosphere during the major deceleration, with assumptions:

1. A constant path angle.

2. A constant drag coefficient.

3. Gravitational force small compared to drag force.

The variation of velocity and deceleration with altitude is shown in **Figures 3** and **4** for various values drag–weight parameter. In each case, it is seen that similar curves result with relative displacements for various values of this parameter.

One interesting facet of these results is that the major deceleration of a body during atmospheric entry takes place in a stratum of the atmosphere of a constant thickness. The level of this stratum in the atmosphere is, however, dependent on the value of the parameter drag-weight ratio. Large values of this parameter because high deceleration in the atmosphere, while smaller values delay deceleration until the lower atmosphere is reached.

Because the decrease velocity during ballistic entry into the Earth's atmosphere is greater than increase density, therefore, the maximum deceleration occurs. The maximum deceleration during entry is independent of drag–weight characteristics of the body.

It is dependent only on the initial angle of inclination, initial entry velocity, and on the density distribution in the Earth's atmosphere; however, the altitude or altitude density at which the maximum deceleration occurs is dependent upon the weight-drag characteristics of the body. **Figures 5** and **6** illustrate the effect of entry angle on the variation of velocity and deceleration during penetration of the Earth's atmosphere from space by a body having the weight-drag characteristics.

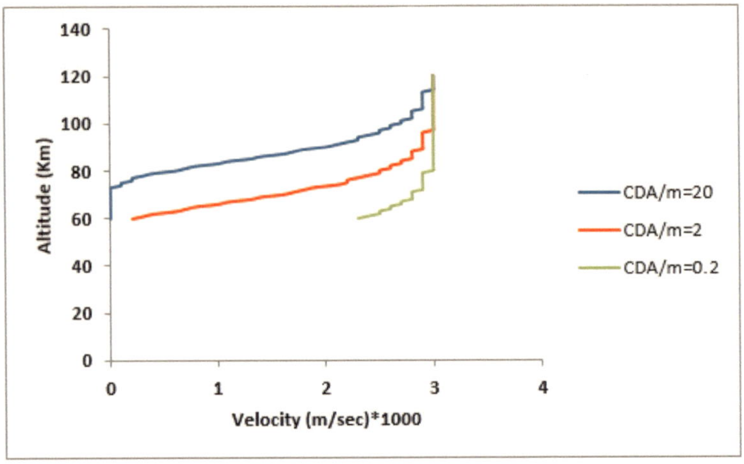

Figure 3.
Variation of velocity with altitude for various values of drag-weight parameter.

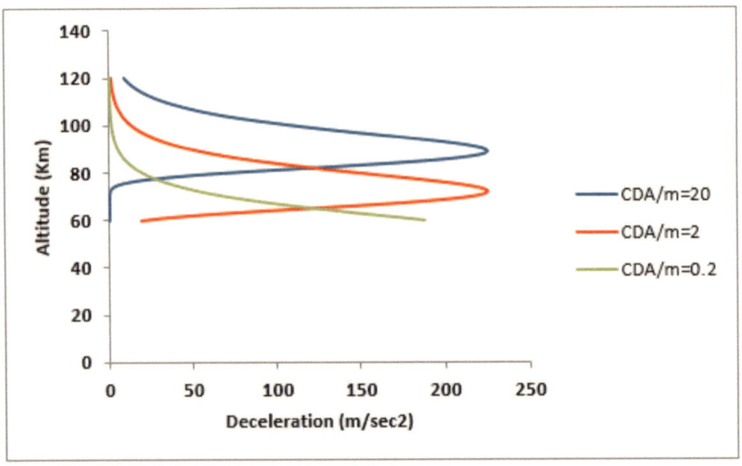

Figure 4.
Variation of deceleration with altitude for various values of drag-weight parameter.

The initial angle of inclination has no effect on the shape of the curve velocity, but on the altitude at which the decrease velocity occurs.

The small values for initial angle of inclination cause decrease in the velocity in higher atmosphere, while the large values cause decrease in the velocity in lower atmosphere. This effect is shown in **Figure 5**.

Figure 6 illustrates variation of deceleration with altitude for various values of the initial angle of inclination. The smaller initial angle of inclination is seen to result in a deceleration higher in the atmosphere and a smaller peak deceleration.

Figure 7 illustrates the variation velocity with altitude for ballistic entry into the Earth's atmosphere with various values of initial entry velocity. We note from this figure that the initial entry velocity has no effects on the shape of the velocity curves.

The initial entry velocity effects the maximum deceleration for the large values for initial entry velocity generate high peak deceleration, while the small values generate low peak deceleration. These effects are shown in **Figure 8**.

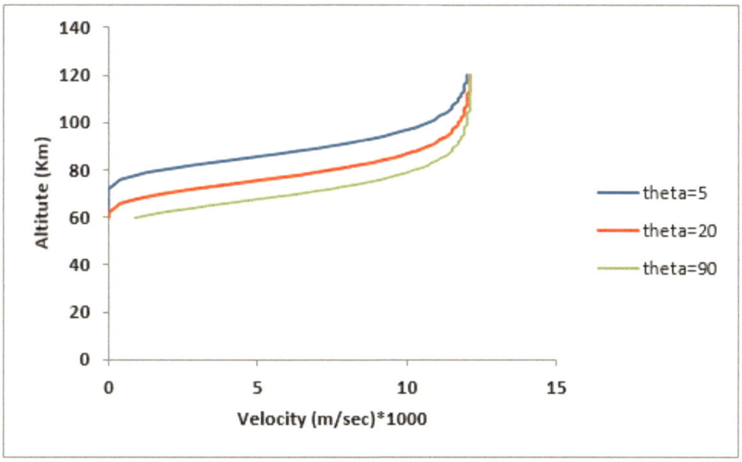

Figure 5.
Variation of velocity with altitude for various values of initial angle of inclination.

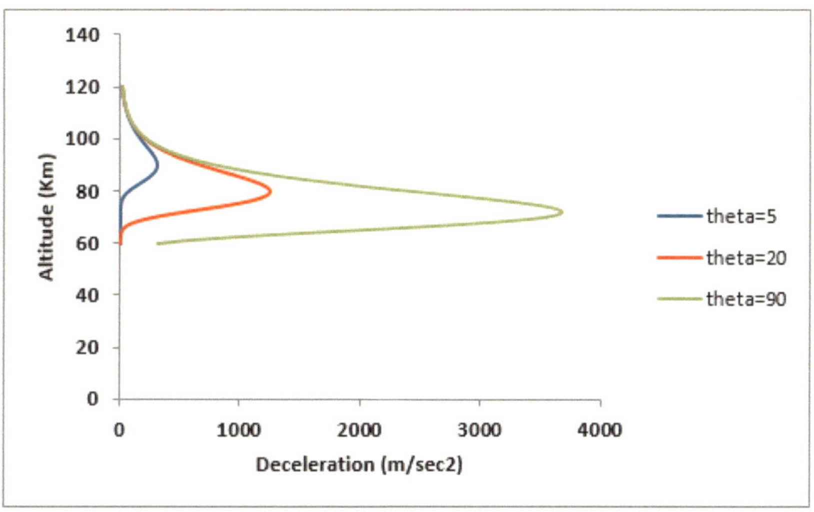

Figure 6.
Variation of deceleration with altitude for various values of initial angle of inclination.

It is interesting to note that these entry solutions are valid not only for relatively steep atmospheric entries from space at escape velocity or near orbital velocity but also for re-entry phase of long-range ballistic trajectories. In this case, the angle of inclination is also relatively constant over the critical parts of the re-entry portion of the trajectory.

5. Conclusion

When anybody enters the atmosphere, a friction process occurs with the particles that make up the atmosphere, and this depends on the cross-sectional area of the body, as the pressure is inversely proportional to the cross-sectional area. The larger

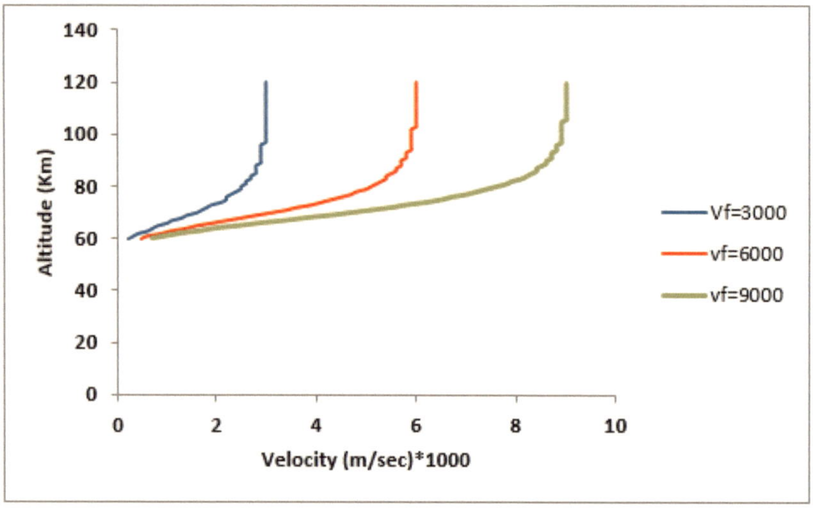

Figure 7.
Variation of velocity with altitude for various values of initial velocity.

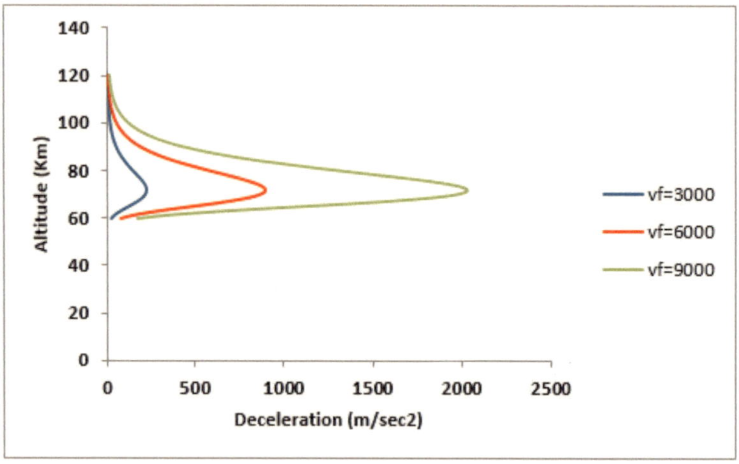

Figure 8.
Variation of deceleration with altitude for various values of initial velocity.

the cross-sectional area, the greater the pressure, and vice versa. When the pressure increases, the process of friction with the particles of the atmosphere will increase, and thus, high heat will be generated, which can cause damage to the inner body. This also depends on the angle of entry, as well as the speed of entry and the density of the atmosphere, as the friction process decreases at entry, because the density is low, and when reaching low altitudes, friction increases due to the high density of particles that make up the atmosphere.

Author details

Omar A. Fadhil
Department of Astronomy and Space-Science of College, University of Baghdad, Baghdad, Iraq

*Address all correspondence to: om19ar95@gmail.com

References

[1] Johnson E, Kannan S. Guidance, Navigation, And Control Conference and Exhibit, Guidance, Navigation, and Control and Co-Located Conference, AIAA Paper 2002–4457. Calif, USA: Monterey; 2002

[2] Yang L, Chen W, Liu X, Zhou H. Steady glide dynamic modeling and trajectory optimization for high lift-to-drag ratio reentry vehicle. International Journal of Aerospace Engineering. Hindawi Publishing Corporation; 2016; **2016**:14. DOI: 10.1155/2016/3527460

[3] Regan FJ, Anandakrishnan SM. Dynamics of Atmospheric Re-Entry. Washington, DC: AIAA Education Series, American Institute of Aeronautics and Astronautics; 1993

[4] Regan FJ. Re-Entry Vehicle Dynamics. Washington, DC: AIAA Education Series, American Institute of Aeronautics and Astronautics; 1984

[5] Bertin JJ, Aerothermodynamics H. AIAA Education Series. Washington, DC: American Institute of Aeronautics and Astronautics; 1994

[6] Hankey WL, Aerodynamics R-E. AIAA Education Series. Washington, DC: American Institute of Aeronautics and Astronautics; 1988

[7] Ashley H. Engineering Analysis of Flight Vehicles. Reading, MA: Addison-Wesley Publishing Company; 1974

Chapter 2

Method for GPS-Monitoring of Small-Scale Fluctuations of the Total Electron Content of the Ionosphere for Predicting the Noise Immunity of Satellite Communications

Vladimir Pashintsev, Mark Peskov, Dmitry Mikhailov, Mikhail Senokosov and Dmitry Solomonov

Abstract

The chapter is devoted to the development of a method for monitoring small-scale fluctuations in the total electronic content of the ionosphere using signals from global navigation satellite systems GPS/GLONASS and for evaluating the characteristics of such fluctuations in the interests of analyzing the noise immunity of satellite communication systems. The proposed method is based on digital processing of measurement results obtained using the advanced capabilities of the NovAtel GPStation-6 receiver and allows to estimate the root mean square of small-scale fluctuations in the total electron content of the ionosphere, the ionospheric scintillation index and to predict the changes in the noise immunity of satellite communication systems.

Keywords: satellite communication systems, global navigation satellite systems, noise immunity, total electron content, small-scale fluctuations, scintillation, digital filter

1. Introduction

It is known [1–9] that the propagation of radio waves in satellite systems under conditions of the formation of intense small-scale irregularities in the Earth's ionosphere is accompanied by the appearance of ionospheric scintillation (fading) of received signals. These effects can lead to a significant decrease in the noise immunity of satellite communication systems (SCS) because of the occurrence of errors when receiving messages. Therefore it is necessary to monitor small-scale irregularities of the ionosphere and to predict changes in the noise immunity of SCS during ionospheric disturbances.

To measure the characteristics of large-scale irregularities of the ionosphere, GPS monitoring methods are most widely used. GPS monitoring is based on measuring the total electron content (TEC) in a radio line of global navigation satellite systems (GNSS) using a dual-frequency receiver. However, ionospheric scintillation is generated not by large-scale fluctuations of the ionospheric TEC, but by its small-scale fluctuations, which are not currently measured by a dual-frequency GNSS receiver. Therefore, the task of modifying a dual-frequency GNSS receiver in the direction of separating (filtering) small-scale TEC fluctuations from fluctuations of larger scales and subsequent determination of their statistical characteristics is very actual.

The aim of the work is to develop a method for GPS monitoring of small-scale fluctuations of ionospheric TEC and their use for predicting changes in the noise immunity of SCS during ionospheric disturbances.

2. Analysis of the effect of small-scale ionospheric irregularities on the noise immunity of satellite communication system

It is known [2–5, 8–12] that the Earth's ionosphere is constantly exposed to natural (solar activity, hurricanes, earthquakes, etc.) disturbing factors that cause change in its electron concentration $N(h,\rho)$ (m^{-3}) in height (h) and space ($\rho = x,y$). This change manifests itself in the formation of vast (up to several hundred and thousands of kilometers) regions (**Figure 1**), in which the electron concentration $N(\rho,h) = \overline{N}(h) + \Delta N(\rho,h)$ differs from the average value $\overline{N}(h)$ due to the formation of irregularities $\Delta N(\rho,h)$ of various spatial scales (from hundreds of kilometers to tens of meters).

The cause of ionospheric scintillation is small-scale irregularities of electron concentration $\Delta N(\rho,h)$, which are characterized by sizes $l_s \approx 10...1000$ m. The magnitude of relative fluctuations ($\Delta N/\overline{N}$) of the electron concentration in small-scale

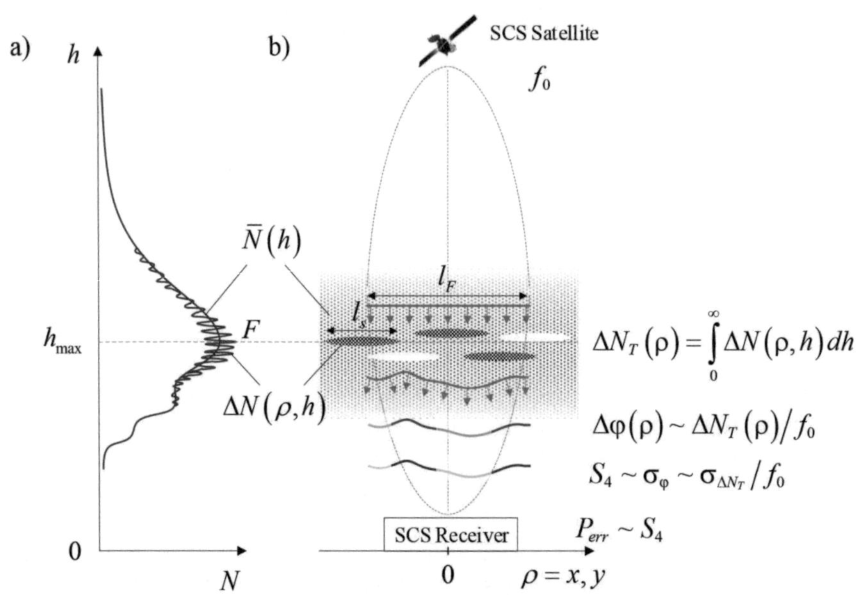

Figure 1.
A model of the ionosphere with small-scale irregularities (a) and trans-ionospheric radio wave propagation (b).

irregularities at any height (h) is practically constant [2, 5, 6, 10]: $\Delta N/\overline{N} \approx \Delta N(\rho, h)/\overline{N}(h) \approx 10^{-3}$. Therefore, the greatest fluctuations of the electron concentration will be observed (**Figure 1**) at the heights of the maximum ionized layer F ($h_{\max} \approx 250...350$ km) of the ionosphere: $\Delta N(\rho, h_{\max}) \approx 10^{-3}\overline{N}(h_{\max})$. With natural and artificial disturbances of the ionosphere, these fluctuations can increase by 1–2 orders of magnitude [2–4, 8, 10] up to $\Delta N(\rho, h_{\max}) \approx (10^{-2}...10^{-1})\overline{N}(h_{\max})$. This makes it possible to represent a set of small-scale irregularities of the ionosphere on the path of propagation of radio waves from the satellite to the receiver in the form of a thin layer (phase screen), which is described by spatial small-scale TEC fluctuations [2, 8]

$$\Delta N_T(\rho) = \int_0^{\infty} \Delta N(\rho, h) dh \tag{1}$$

relative to its average value $\overline{N}_T = \int_0^{\infty} \overline{N}(h) dh$ (m^{-2}). According to Eq. (1), with ionospheric disturbances, small-scale TEC fluctuations can increase by 1–2 orders of magnitude.

According to [2, 7, 8], the process of radio wave propagation (**Figure 1**) with a carrier frequency f_0 from the SCS satellite to the receiver through an ionospheric layer with small-scale irregularities (phase screen) is accompanied by distortions (fluctuations) of its phase front within the region of space ρ bounded by the diameter of the first Fresnel zone ($l_F \approx 2\sqrt{ch_{\max}/f_0} \geq l_s$):

$$\Delta\varphi(\rho) = -80,8\pi\Delta N_T(\rho)/cf_0, \text{rad}, \tag{2}$$

where c – the speed of light (m/s); 80,8 – coefficient having dimension m^3/s^2.

The interference of individual sections (ρ) of the phase front of the wave Eq. (2) behind the phase screen causes a redistribution of the amplitude $A(\rho, h)$ along the wave front (illustrated in **Figure 1** in the form of a change in its thickness). These diffraction processes determine the instantaneous value of the signal power $P_r = A^2$ at the receiver input. Therefore, a change in the spatial TEC fluctuations $\Delta N_T(\rho)$ along the wave propagation path due to the drift of small-scale irregularities and the motion of the satellite leads to a random change in the power of received signal carrier (P_r)—fading or scintillation. It leads to change in the ratio $P_r/N_0 \equiv C/N_0$ of signal power to the spectral power density of noise (N_0) at the receiver input (carrier-to-noise ratio). To assess the degree of manifestation of this scintillation, the value of the scintillation index is traditionally used [1–4, 9, 13]

$$S_4 = \sqrt{\left(\langle P_r^2\rangle - \langle P_r\rangle^2\right)/\langle P_r\rangle^2}, \tag{3}$$

where $\langle x\rangle$—designation of statistical averaging of random values x.

As an illustration, **Figure 2** shows the changes in the carrier-to-noise ratio (C/N_0) and the scintillation index (S_4) of the GNSS GLONASS signal in the L1 frequency range obtained using the NovAtel GPStation-6 receiver, which is located at the North-Caucasus Federal University (Stavropol, Russia).

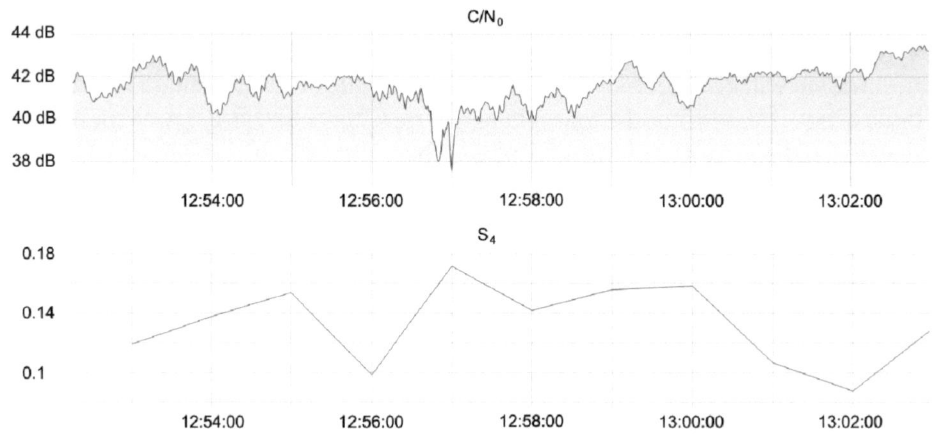

Figure 2.
Changes in the carrier-to-noise ratio (top panel) and the scintillation index (bottom panel) of the GNSS GLONASS signal in the L1 frequency range.

Analysis of **Figure 2** shows that ionospheric scintillation is manifested around 12:56:40 local time (UTC + 3) by a sharp decrease in the carrier-to-noise ratio by 2 ... 3 dB for about 20 seconds. This leads to a short-term increase in the scintillation index by almost 2 times (from $S_4 \approx 0.1$ to $S_4 \approx 0.17$). According to [1–4, 8, 9, 14] in natural conditions, such short-term scintillation with relatively low intensity ($S_4 = 0.01...0.15$) is characteristic of mid-latitude regions. In the polar regions, the scintillation is more intense ($S_4 = 0.45...0.7$) and lasts from a few seconds to several minutes. In the equatorial regions, scintillation can last from several minutes to several hours and the scintillation index can reach $S_4 = 0.9...1$. It is obvious that such significant differences in the intensity and duration of scintillation depend on the degree of local changes in the ionosphere electron concentration and the causes of their occurrence.

In the absence of small-scale fluctuations of the ionosphere electron concentration and TEC ($\Delta N_T(\rho) \sim \Delta N(\rho, h) \to 0$) phase fluctuations in the wave front at the output of the ionosphere Eq. (2) will be absent ($\Delta\varphi(\rho) \sim \Delta N_T(\rho) \to 0$). It causes the absence of diffraction effects and scintillation of the received signals ($S_4 \to 0$). As the small-scale TEC fluctuations ($\Delta N_T(\rho)$) increase or carrier frequency (f_0) decrease, the phase fluctuations in the wave front $\Delta\varphi(\rho) \sim \Delta N_T(\rho)/f_0$ will increase proportionally. It follows that the ionospheric scintillation index ($S_4 \sim \sigma_\varphi$) will increase with increasing root mean square (RMS) $\sigma_\varphi = \langle \Delta\varphi^2(\rho) \rangle^{1/2}$ of the phase front fluctuations at the ionospheric output (phase screen).

For the case when the fluctuations of the phase front $\Delta\varphi(\rho)$ at the output of the phase screen are described by a Gaussian probability distribution with zero mathematical expectation and dispersion σ_φ^2, the dependence $S_4 = \psi(\sigma_\varphi)$ has the form [13]

$$S_4 = \sqrt{1 - \exp\left(-2\sigma_\varphi^2\right)}. \tag{4}$$

Taking into account Eq. (2) the RMS of the fluctuations of the phase front at the ionosphere output (σ_φ) is determined by the RMS of small-scale TEC fluctuations $\sigma_{\Delta N_T} = \langle \Delta N_T^2(\rho) \rangle^{1/2}$ and the carrier frequency f_0 [2, 8]:

$$\sigma_\varphi = 80,8\pi\sigma_{\Delta N_T}/cf_0. \tag{5}$$

Thus, in accordance with Eqs. (4) and (5), the following dependence $S_4 = \psi(\sigma_{\Delta N_T})$ of the ionospheric scintillation index on the small-scale TEC fluctuations takes place [9]:

$$S_4 = \sqrt{1 - \exp\left(-2\left(80,8\pi\sigma_{\Delta N_T}/cf_0\right)^2\right)}. \tag{6}$$

The analysis showed that, in accordance with Eqs. (3)–(6), as the small-scale TEC fluctuations ($\sigma_{\Delta N_T}$) and the phase front fluctuations of the output wave ($\sigma_\varphi \sim \sigma_{\Delta N_T}/f_0$) increase, the ionospheric scintillation index ($S_4 \sim \sigma_\varphi \sim \sigma_{\Delta N_T}/f_0$) increases. Under these conditions, the quality of SCS functioning can significantly decrease due to the occurrence of errors when receiving messages. Signals with offset binary phase shift keying (OBPSK) are most commonly used to transmit messages in SCS. The probability of error (P_{err}) when receiving such signals with a constant average signal-to-noise energy ratio at the receiver input ($\langle h^2\rangle$) under conditions of ionospheric scintillation (characterized by Nakagami distribution) is determined by the expression [15]:

$$P_{err} = \frac{1}{2}\left(\frac{m}{m + \langle h^2\rangle}\right)^m = \frac{1}{2}\left(\frac{S_4{}^{-2}}{S_4{}^{-2} + \langle h^2\rangle}\right)^{S_4{}^{-2}}, \tag{7}$$

where $m = 1/S_4^2$—Nakagami distribution parameter.

The analysis of Eq. (7) taking into account Eqs. (3)–(6) allows us to conclude that as the small-scale TEC fluctuations ($\sigma_{\Delta N_T}$) increase, the ionospheric scintillation index ($S_4 \sim \sigma_{\Delta N_T}/f_0$) and the probability of error ($P_{err} \sim 1/m \sim S_4$) when receiving signals that correspond to the elementary symbols of the transmitted message increase. Therefore, the prediction of its changes during ionospheric disturbances is an actual problem.

It is possible to solve this problem in two ways based on the results of monitoring the scintillation index (S_4):

1. in accordance with Eqs. (3) and (7) based on the results of measuring fluctuations of the received signal power (P_r) in GNSS;

2. in accordance with Eqs. (6) and (7) based on the results of the measurement of small-scale TEC fluctuations ($\sigma_{\Delta N_T}$).

Evaluation of the scintillation index S_4 is currently carried out using specialized dual-frequency GNSS receivers of the GISTM class (GNSS Ionospheric Scintillation and TEC Monitor) [11, 16, 17]. However, it is impractical to use them (including GPStation-6) to estimate the ionospheric scintillation index in the interests of noise immunity prediction for three reasons.

Firstly, the calculation of the scintillation index (S_4) in these receivers is carried out according to Eq. (3) based on the measurement of fluctuations of the received signal power (P_r). At low elevation angles of the GNSS satellite ($\alpha < 5°...15°$), scintillation can be caused not only by small-scale ionospheric irregularities but also by multiple reflection of radio waves from a variety of objects surrounding the receiving antenna. Therefore, the reliability of the calculation of the scintillation index, in this case, will be lower than on the basis of the estimation of small-scale fluctuations of the ionospheric TEC (Eq. (6)).

Secondly, the sampling interval of the results of calculating the scintillation index (S_4) in modern receivers usually does not exceed 1 minute. This means that its current value does not allow detection of short-term (lasting up to several seconds) occurrence of scintillation and reliably assess their intensity. This is confirmed by **Figure 2**.

Thirdly, the carrier frequency (f_0) in SCS may differ significantly from the carrier frequency of GNSS, so the scintillation indices ($S_4 \sim \sigma_{\Delta N_T}/f_0$) in different systems will also differ.

On the other hand, it is not possible to carry out an assessment of the RMS of small-scale fluctuations ($\sigma_{\Delta N_T} = \langle \Delta N_T^2(\rho) \rangle^{1/2}$) based on the results of measuring the ionospheric TEC (N_T) using the GPStation-6 receiver for a number of reasons related to the complexity of separating (filtering) of small-scale TEC fluctuations (ΔN_T) with dimensions $\rho \approx l_s = 10...1000$ m from large- and medium-scale TEC fluctuations with dimensions $\rho > l_s = 1000$ m.

Therefore, the aim of the work is to develop a method for GPS monitoring of small-scale fluctuations of ionospheric TEC (ΔN_T) and their use for predicting changes in the noise immunity of SCS ($P_{err} \sim S_4 \sim \sigma_{\Delta N_T}/f_0$) during ionospheric disturbances. To achieve it first of all it is necessary to modify the GISTM-receiver for measuring small-scale TEC fluctuations (ΔN_T). Further, on this basis, it is possible to obtain estimates of the RMS of small-scale TEC fluctuations ($\sigma_{\Delta N_T}$), the ionospheric scintillation index in the SCS ($S_4 \sim \sigma_{\Delta N_T}/f_0$) according to Eq. (6) and to predict the changes in the noise immunity of the SCS ($P_{err} = \psi(S_4, \langle h^2 \rangle)$) according to Eq. (7).

3. Modification of the GPStation-6 receiver for measuring small-scale fluctuations of the total electron content of the ionosphere

It is known [11] that the method of GPS monitoring of the ionosphere based on the transmission of signals from the GNSS satellite at two carrier frequencies (f_1 and f_2) and the use of dual-frequency GNSS receivers allows to measure the values of the parameters of the received signals with high accuracy and on their basis to calculate not only the scintillation index (S_4) but also the ionospheric TEC (N_T).

Figure 3 shows a simplified structure for the construction of a dual-frequency receiver GPStation-6 for code measurements of ionospheric TEC (TEC_c) and its modification for the evaluation of small-scale TEC fluctuations (ΔN_T). The receiver GPStation-6 consists of 2 main parts [17, 18]:

1. the hardware part includes a signal reception unit that performs analog-to-digital conversion, decoding of GNSS signals, and measurement of their main parameters: pseudo-ranges (R_1', R_2') to satellite and pseudo-phases (φ_1', φ_2');

2. the software part implementing a digital processing unit in which the calculation of radio navigation parameters, receiver coordinates, scintillation index (S_4), ionospheric TEC (TEC_c), etc., is carried out.

In order to estimate small-scale fluctuations of the ionospheric TEC (ΔN_T) based on the results of measuring the ionospheric TEC (N_T) using a dual-frequency receiver GPStation-6, it is necessary to modify its software (**Figure 3**).

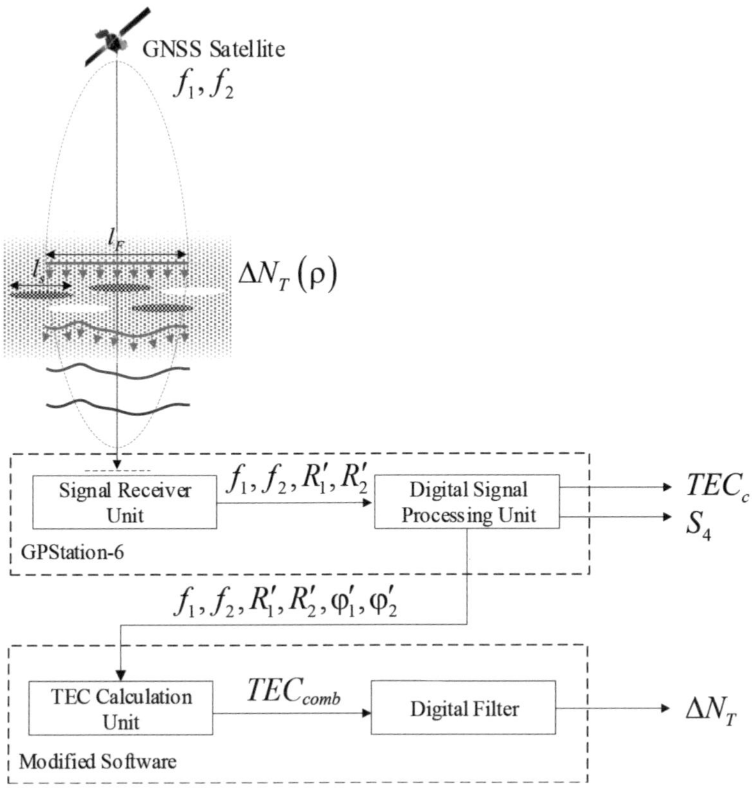

Figure 3.
The structure of the construction of the GPStation-6 dual-frequency receiver for code measurements of ionospheric TEC (TEC_c) and its modification for the evaluation of small-scale TEC fluctuations (ΔN_T).

The need for such a modification of the GPStation-6 receiver software is due to the following reasons. The principle of operation of the GISTM receiver is based on the results of measurements of the pseudo-range $R_i' \sim R + 40,4N_T/f_i^2$ to each of the navigation satellite (otherwise, code measurements). In addition to the true range R, they contain an ionospheric component that depends on the ionospheric TEC (N_T) on the route from the satellite to the receiver and the carrier frequency (f_i) [11, 16, 17]. Since the transmission of navigation signals in the GNSS is carried out simultaneously at two carrier frequencies ($f_1 \approx 1,6$ GHz and $f_2 \approx 1,2$ GHz), it is possible to calculate the ionospheric TEC at each moment of time based on a linear combination (difference) of the measured pseudo-ranges ($R_1' \sim R + N_T/f_1^2$ and $R_2' \sim R + N_T/f_2^2$) [11, 19]:

$$
\begin{aligned}
TEC_c &= \frac{1}{40,4}\left(\frac{f_1^2 f_2^2}{f_1^2 - f_2^2}\right)(R_2' - R_1') \\
&= \frac{1}{40,4}\left(\frac{f_1^2 f_2^2}{f_1^2 - f_2^2}\right)(R_2 - R_1) + \delta_d + \delta_n = N_T + \delta_d + \delta_n.
\end{aligned}
\tag{8}
$$

According to Eq. (8), the results of the code measurements of the ionospheric TEC (TEC_c) differ from its true value (N_T) by the amount of systematic error δ_d due to differential signal delays in transmitting and receiving radio paths, and random error δ_n due to the influence of internal receiver noise.

Code measurements of the TEC (TEC_c) are made in the GPStation-6 receiver with a sampling interval $\tau_d = 1$ s. Therefore, as discrete values accumulate, a time series is formed

$$TEC_c(t) = N_T(t) + \delta_d + \delta_n. \tag{9}$$

In the general case of slant propagation of radio waves (at an angle α relative to the horizontal) through the ionosphere with electron concentration irregularities $N(\rho, h) = \overline{N}(h) + \Delta N(\rho, h)$, the measured TEC, taking into account Eq. (1), is described as the sum of the regular (average) and fluctuation components:

$$N_T(\rho) = \int_0^\infty N(\rho, h)dz = \int_0^\infty \overline{N}(h)dz + \int_0^\infty \Delta N(\rho, h)dz = \overline{N}_T + \Delta N_T(\rho), \tag{10}$$

where $dz \approx dh$ cosec α—an elementary section of the slant path of radio waves propagation.

In this case, the results of code measurements of the TEC of the inhomogeneous ionosphere (Eq. (10)) will be described by the equation

$$TEC_c(t) = N_T(t) + \delta_d + \delta_n \approx \overline{N_T}(t) + \Delta N_T(t) + \delta_d + \delta_n. \tag{11}$$

The value of the systematic error $\delta_d = \delta_{\text{SDCB}} + \delta_{\text{RDCB}}$ consists of the differential signal delays in the transmitting (δ_{SDCB}) and receiving (δ_{RDCB}) radio paths. In this case, the value of δ_{RDCB} can be obtained from ISMCALIBRATIONSTATUS logs at the output of the GPStation-6 receiver as a result of performing the automatic calibration procedure [18]. The values of δ_{SDCB} are published daily on the web portal ftp://ftp.unibe.ch/aiub/CODE. Therefore, the value of systematic error $\delta_d = \delta_{\text{SDCB}} + \delta_{\text{RDCB}}$ can be calculated and excluded from the measurement results ($\delta_d \approx 0$).

The magnitude of the relative fluctuations of the electron concentration in the small-scale irregularities of the ionosphere is $\Delta N/\overline{N} \approx \Delta N(\rho, h)/\overline{N}(h) \approx 10^{-3}$, therefore, the relative small-scale TEC fluctuations have the same order: $\Delta N_T/\overline{N}_T \approx 10^{-3}$ or $\Delta N_T \approx 10^{-3}\overline{N}_T$. The value of the random (noise) error of the code measurements of the ionospheric TEC can reach 50% of its average value: $\delta_n \approx 0,5\overline{N}_T$ [11]. Therefore, when using such measurements, a ratio $\Delta N_T(t) < < \delta_n$ is carried out that indicates the "noise" of the results of the assessment of small-scale TEC fluctuations. Then Eq. (11) takes the form $TEC_c(t) \approx \overline{N}_T(t) + \delta_n$.

To eliminate "noise", it is necessary to modify the software of the GPStation-6 receiver by replacing code measurements with a combination of code and phase measurements [11, 19]. To do this, the software (**Figure 2**) includes a unit for calculating the TEC by code-phase measurements. When implementing combined (code-phase) measurements of the TEC, its value at each moment of time is calculated based on the results of pseudo-ranges (R_1', R_2') and pseudo-phases (φ_1', φ_2') measurement according to the eq. [11, 19]

$$TEC_{comb} = \frac{1}{40,4}\left(\frac{f_1^2 f_2^2}{f_1^2 - f_2^2}\right)(\lambda_2\varphi_2' - \lambda_1\varphi_1') - \delta_a, \tag{12}$$

where $\lambda_i = c/f_i$—the wavelength at the appropriate frequency (f_1, f_2); $\delta_a = \psi(f_1, f_2, R'_1, R'_2, \varphi'_1, \varphi'_2, \alpha)$—correction to resolve the ambiguity of phase measurements (characterizes the average offset of the results of the TEC calculation based on ambiguous phase measurements TEC_p relative to its true value: $\delta_a = \overline{N_T} - \overline{TEC_p}$).

In this case, the noise error of measuring the ionospheric TEC is practically eliminated ($\delta_n \approx 0$) and the results of combined measurements of the TEC at the output of the TEC calculation unit for code-phase measurements will be described by the sum $TEC_{comb}(t) \approx \overline{N}_T(t) + \Delta N_T(t)$ of the average value of the ionospheric TEC and its small-scale fluctuations.

4. Estimation of small-scale fluctuations of the total electron content of the ionosphere

4.1 Characteristics of small-scale fluctuations of the total electron content of the ionosphere

It should be noted that the average value of the ionospheric TEC generally includes medium-scale ($\Delta N_{T(med)}(\rho)$) and large-scale ($\Delta N_{T(lrg)}(\rho)$) TEC fluctuations (variations) relative to the background ($\overline{N}_{T0}$) value: $\overline{N}_T(\rho) = \overline{N}_{T0} + \Delta N_{T(med)}(\rho) + \Delta N_{T(lrg)}(\rho)$. According to [11], medium-scale fluctuations of TEC ($\Delta N_{T(med)}(\rho)$) have dimensions $\rho = 50...300$ km, large-scale TEC fluctuations have dimensions $\rho > 1000$ km, and they do not cause ionospheric scintillation. Taking into account the irregularities of different scales, the ionospheric TEC (Eq. (10)) will be described by an equation of the general form

$$N_T(\rho) = \overline{N}_{T0} + \Delta N_{T(med)}(\rho) + \Delta N_{T(lrg)}(\rho) + \Delta N_T(\rho), \quad (13)$$

according to which a time series will be formed at the output of the TEC calculation unit based on code-phase measurements

$$TEC_{comb} = \overline{N}_T(t) + \Delta N_T(t) = \overline{N}_{T0} + \Delta N_{T(med)}(t) + \Delta N_{T(lrg)}(t) + \Delta N_T(t). \quad (14)$$

In order to isolate (filter) small-scale fluctuations $\Delta N_T(t)$ from the time series of TEC by code-phase measurements (Eq. (14)) and preliminary substantiation of the parameters of the digital filter (**Figure 3**), it is necessary first of all to establish the dependence TEC_{comb} on the spatial position of the irregularities $\Delta N(\rho, h)$ of the electron concentration of the ionosphere with scales $\rho \approx 10...1000$ m. **Figure 4** shows a method for estimating the time characteristics of small-scale fluctuations of the ionospheric TEC using a GPStation-6 receiver based on results of TEC calculation at time t_1 and t_2.

At the moment t_1 the radio wave propagation path from the satellite to the receiver intersects the starting point (ρ_1, h) of a small-scale irregularity of the ionosphere with an electron concentration $N(\rho_1, h) = \overline{N}(h) + \Delta N(\rho_1, h)$ and an average size l_s. In accordance with Eq. (10), Eq. (13) and **Figure 4**, this value of the electron concentration corresponds to the TEC

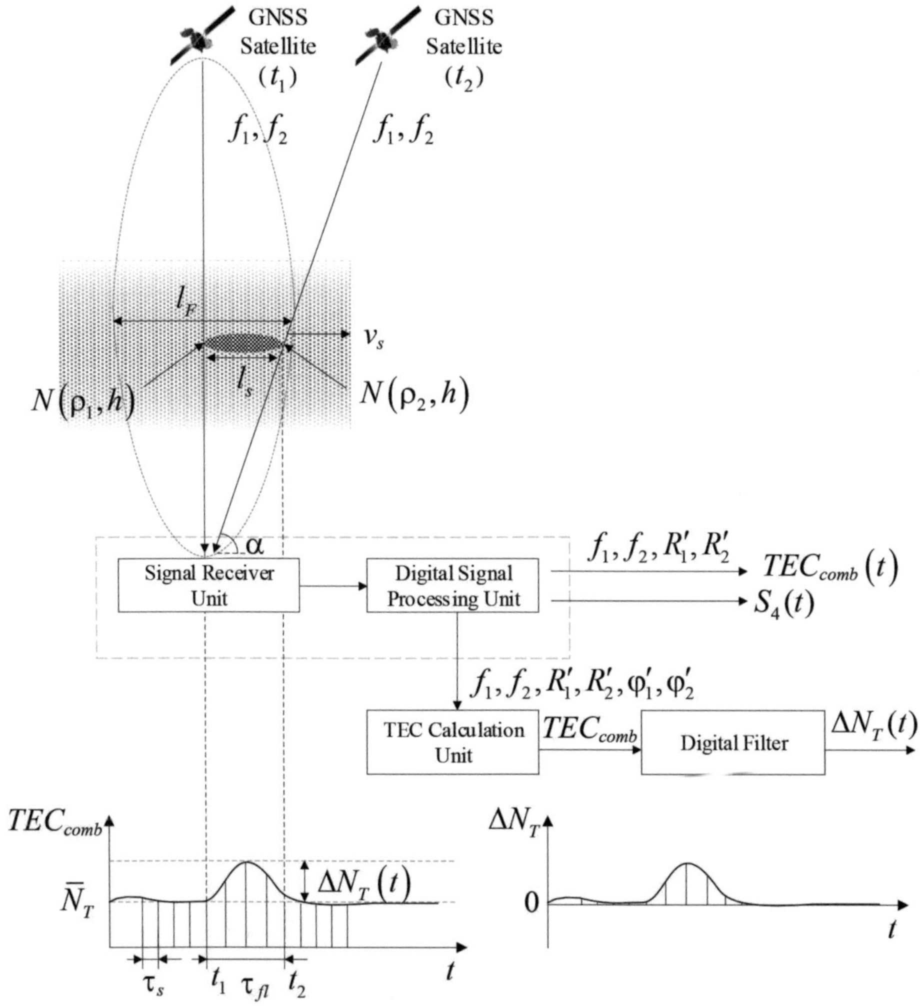

Figure 4.
A method for measuring the time characteristics of small-scale fluctuations of the ionospheric TEC using the GPStation-6 receiver.

$$N_T(\rho_1) = \int_0^\infty \overline{N}(h)dz + \int_0^\infty \Delta N(\rho_1, h)dz = \overline{N}_T + \Delta N_T(\rho_1). \tag{15}$$

Therefore, at a time close to t_1, at the output of the receiver and the TEC calculation unit for code-phase measurements, the result of measuring the TEC will be formed as $TEC_{comb}(t_1) \approx \overline{N}_T(t_1) + \Delta N_T(t_1)$.

The usual [3] drift velocity of small-scale irregularities in the ionosphere layer F ($v_d \approx 100...150$ m/s) is much less than the velocity of the satellite ($v_{satellite} \approx 7,9$ km/s) and its projection to the height of the ionosphere layer F ($h_{\max} \approx 250...350$ км). In [3] it is called the effective scanning velocity and is defined as

$$v_s = 2\pi(R_{\text{Earth}} + h_{\max})/T_{satellite} \approx 1000 \text{ m/s} \gg v_d, \tag{16}$$

where $R_{Earth} = 6371$ km—the radius of the Earth; $T_{satellite} \approx 40\ 000$ s—the average nodal period of the GNSS GPS/GLONASS satellite. Therefore, it can be assumed that the irregularity of the electron concentration will not change its position (**Figure 3**) during the time $\Delta t = t_2 - t_1$ of movement of the radio wave propagation path at a distance l_s to the endpoint (ρ_2, h) of the small-scale irregularity with the electron concentration $N(\rho_2, h) = \overline{N}(h) + \Delta N(\rho_2, h)$. In accordance with Eq. (10), Eq. (13) and **Figure 4**, this electron concentration value corresponds to the TEC $N_T(\rho_2) = \overline{N}_T + \Delta N_T(\rho_2)$ on the route from the satellite to the receiver and at a time close to t_2 the TEC measurement result $TEC_{comb}(t_2) \approx \overline{N}_T(t_2) + \Delta N_T(t_2)$ will be generated at the output of the TEC calculation unit. Since the average (background) value of the electron concentration in the ionosphere is assumed to be unchanged ($\overline{N}(h) = const$), the average value of the TEC at t_1 and t_2 does not change: $\overline{N}_T(t_1) = \overline{N}_T(t_2)$. Therefore, small-scale TEC deviations (fluctuations) at the output of the receiver and the TEC calculation unit by code-phase measurements

$$TEC_{comb}(t) \sim \Delta N_T(t) \sim \Delta N_T(\rho) \sim \Delta N(\rho, h) \tag{17}$$

will be observed at a time interval $\tau_{fl} = \Delta t = t_2 - t_1$ that corresponds to the displacement of the intersection area of the radio wave propagation path at a distance $\Delta\rho = \rho_2 - \rho_1 = l_s$ equal to the average size of small-scale irregularity at a velocity of v_s.

The influence of the spatial characteristics (average size l_s) of small-scale ionospheric irregularities on the time characteristics of measured small-scale TEC fluctuations $\Delta N_T(t)$ is described by an equation for estimating the period of these fluctuations:

$$\tau_{fl} = \Delta t = l_s / v_s. \tag{18}$$

The specified characteristic of the TEC fluctuations should more accurately be called the average value of the TEC fluctuations period, since it is determined by the average (characteristic) size l_s of the irregularities. However, the range of changes in small-scale irregularities that cause the appearance of fading (scintillation) of received signals extends from the lowest values ($l_{s\,\min} < l_s$) to the maximum corresponding to the diameter of the Fresnel zone $l_{s\,\max} = l_F \approx 2\sqrt{ch_{\max} cosec\alpha / f_0} \geq l_s$. Therefore, the period of small-scale TEC fluctuations with the largest dimensions $l_s = l_{s\,\max} = l_F$ is defined (by analogy with the average period $\tau_{fl} = \Delta t = l_s / v_s$) as

$$\tau_{fl\ \max} = l_F / v_c. \tag{19}$$

The smallest sizes of small-scale ionospheric irregularities ($l_{s\,\min} < l_s$) can be tens of meters [2]. However, the smallest sizes of small-scale irregularities that can be measured using the GPStation-6 receiver depend on the data sampling period (τ_s). According to the Nyquist relation, the minimum period of fluctuations ($\tau_{fl\ \min}$) that can be recorded with the data sampling period τ_s is $2\tau_s$. It is assumed that the oscillation has a sinusoidal (wave) character. According to [12], in practice, it is not enough to register fluctuations of two samples for a period, since in pure form harmonic disturbances in the ionosphere are not realized. The experience of ionospheric measurements has shown that for the effective measurement of TEC fluctuations from 5 to 10 measurements per period ($5\tau_s ... 10\tau_s$) are required. Therefore, the period of small-scale TEC fluctuations with the smallest dimensions $l_s = l_{s\,\min} = 5\tau_s v_s$ is defined as

$$\tau_{fl\ \min} = l_{s\ \min}/v_s = 5\tau_s. \tag{20}$$

Thus, in order to evaluate the statistical characteristics of small-scale TEC fluctuations (ΔN_T) it is necessary to modify the software of the GPStation-6 receiver (**Figure 3**) in the direction of replacing the code measurements of the TEC (Eq. (11)) with code-phase measurements (Eq. (12)), which eliminates their "noise". Then it is necessary to filter out small-scale fluctuations from the time series of TEC (Eq. (14)) on the basis of estimation (**Figure 4**) of the maximum $\tau_{fl\ \max} = l_F/v_c$ and minimum $\tau_{fl\ \min} = l_{s\ \min}/v_s = 5\tau_s$ periods of small-scale TEC fluctuations.

4.2 Digital filter parameters

The expressions obtained above for estimating the minimum and maximum periods of small-scale TEC fluctuations $\Delta N_T(t)$ allow us to determine the main parameters of the digital filter (**Figure 3**).

The maximum size of small-scale ionospheric irregularities corresponding to the diameter of the Fresnel zone $l_{s\max} = l_F \approx 2\sqrt{ch_{\max} cosec\alpha/f_0}$ on SCS routes (at typical values $h_{\max} \approx 250...350$ km, $\alpha = 15°...90°$ and $f \approx 1,6$ GHz) is $l_{s\max} = l_F \approx 550...1000$ m. Therefore, with $l_{s\max} \approx 1000$ m and $v_s \approx 1000$ м/c we will have $\tau_{fl\ \max} = 1$ s. This value determines the minimum frequency of small-scale TEC fluctuations: $f_{fl\ \min} = 1/\tau_{fl\ \max} = 1$ Hz.

In the GPStation-6 receiver, the measurement of the TEC (TEC_c) is carried out with a sampling period $\tau_s = 1$ s [18]. The minimum period of fluctuations ($\tau_{fl\ \min}$) that can be recorded with such data sampling period is $\tau_{fl\ \min} = 5\tau_s = 5$ s. It follows from this that the capabilities of the GPStation-6 receiver are limited to measuring the irregularities with sizes $l_{s\ \min} = 5\tau_s v_s \approx 5000$ m or more (medium- and large-scale irregularities exceeding the size of the Fresnel zone $l_F \approx 1000$ м). However, it is known [12] that in order to study the fine (small-scale) structure of the ionosphere, it is necessary to measure the TEC with a sampling frequency of at least $f_s = 1/\tau_s = 50$ Hz (with an interval $\tau_s = 1/f_s = 0.02$ s).

The analysis of [20] allows us to conclude that the GPStation-6 receiver is capable of measuring the parameters of GPS/GLONASS navigation signals (contained in RANGE log) with a minimum sampling interval $\tau_s = 0.02$ s (frequency $f_д = 50$ Гц). The obtained data can be used to calculate the ionospheric TEC by code-phase measurements in accordance with Eq. (12). It follows from this that the capabilities of the GPStation-6 receiver make it possible to measure the TEC fluctuations with a minimum period $\tau_{fl\ \min} = 5\tau_s = 0.1$ s. This value determines the maximum frequency of small-scale TEC fluctuations: $f_{fl\ \max} = 1/\tau_{fl\ \min} = 10$ Hz.

The results of the analysis have shown the possibilities and ways to improve the GPStation-6 receiver software in the direction of measuring small-scale TEC fluctuations (ΔN_T). The obtained value of the highest realizable sampling frequency ($f_s = 50$ Hz) and the values of the minimum ($f_{fl\ \min} = 1$ Hz) and maximum ($f_{fl\ \max} = 10$ Hz) frequencies of small-scale TEC fluctuations make it possible to evaluate (isolate) it based on the use of a discrete digital filter in the modified GPStation-6 receiver software (**Figure 3**). On the one hand, it should have the magnitude–frequency response close to ideal (in particular, as smooth as possible at transmission frequencies corresponding to the range from $f_{fl\ \min}$ to $f_{fl\ \max}$). On the other hand, it should not introduce a large delay (τ_{delay}) when processing (filtering) measurement results.

The first condition is satisfied by a digital Butterworth filter with attenuation of 3 dB at the cutoff frequencies corresponding to $f_{fl\ \min} = 1\ \text{Hz}$ and $f_{fl\ \max} = 10\ \text{Hz}$, in which the magnitude–frequency response has the form shown in **Figure 5** and depends on the order (p) of the digital filter [21]. To satisfy the second condition, it is necessary to analyze the dependencies (**Figure 6**) of the group delay (τ_{delay}) of the measurement results in the specified filter on the frequency.

The output samples of the considered digital Butterworth filter are determined by a general Eq. (21)

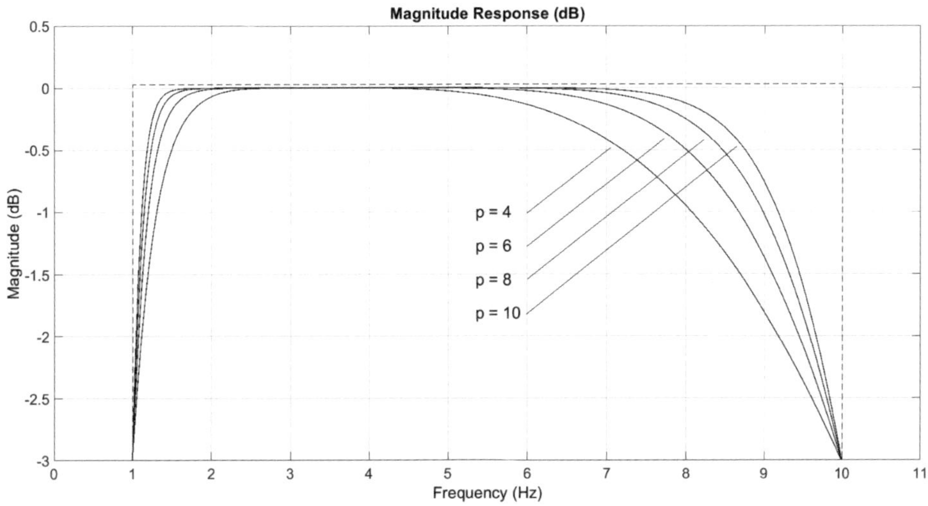

Figure 5.
Magnitude-frequency response of digital Butterworth filters of various orders.

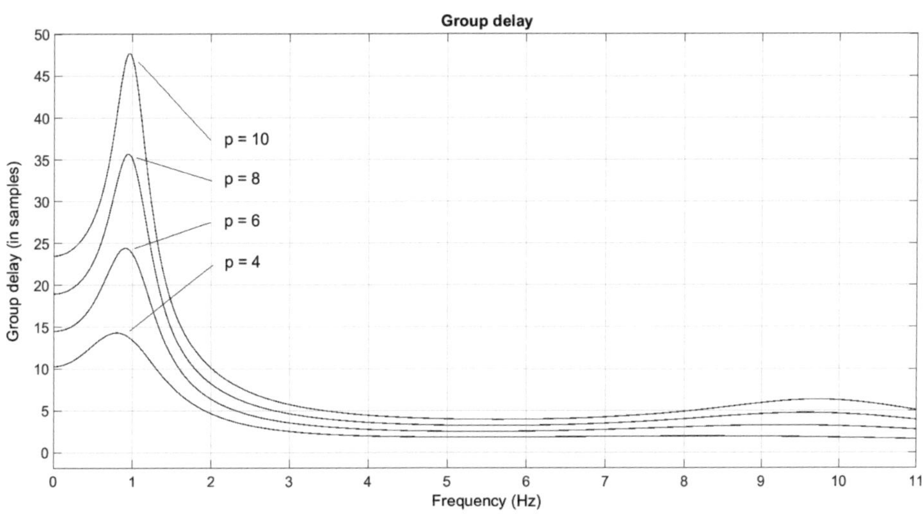

Figure 6.
Dependence of the group delay of digital Butterworth filters of various orders on frequency.

$$y_i = \frac{1}{a_0}\left[\sum_{k=0}^{p} b_k x_{i-k} - \sum_{k=1}^{p} a_k y_{i-k}\right], \tag{21}$$

where the number and values of the coefficients a_k and b_k depend on the sampling frequency (f_s) of the input data (x_i), the filter bandwidth range (limited to values $f_{fl\ \min}$ and $f_{fl\ \max}$) and filter order, which is given by an even integer p. In the case of using this filter to process the results of measuring the ionospheric TEC (TEC_{comb}) in order to evaluate its small-scale fluctuations (ΔN_T), Eq. (21) can be represented as

$$\Delta N_{Ti} = \frac{1}{a_0}\left[\sum_{k=0}^{p} b_k TEC_{comb\ i-k} - \sum_{k=1}^{p} a_k \Delta N_{Ti-k}\right]. \tag{22}$$

To justify the choice of the filter order (p) for measuring small-scale TEC fluctuations, it should be recalled that, on the one hand, it should have the magnitude–frequency response close to ideal, and on the other hand, it should not introduce a large group delay (τ_{delay}) when processing measurement results.

It is known [21] that as the order of the digital Butterworth filter increases, its magnitude-frequency response (**Figure 5**) approaches the ideal (rectangular) simultaneously with the increase in the group delay (**Figure 6**). Let us analyze in more detail the dependencies of the group delay (τ_{delay}) on the frequency in the range from 1 Hz to 5 Hz (**Figure 7**).

On the additional vertical axis in **Figure 7**—the values of the filter delay τ_{delay} in seconds, on the additional horizontal axis – the period of fluctuations τ_{fl} in seconds. As a criterion for the permissible delay introduced by the filter during the processing of measurement results, we will take a value not exceeding half the period of small-scale TEC fluctuations: $\tau_{delay} \leq 0,5\tau_{fl}$. In **Figure 7**, the area of acceptable filter delays is under the dotted curve.

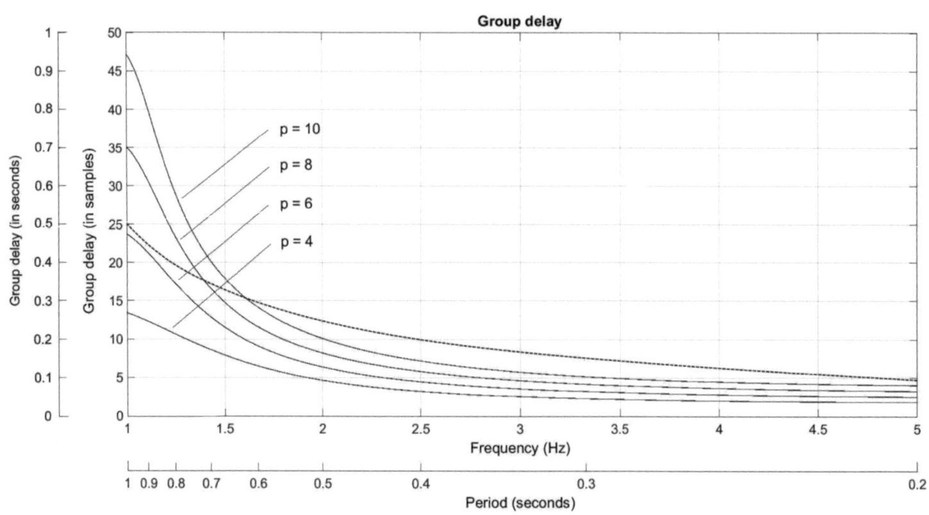

Figure 7.
Dependence of the group delay of digital Butterworth filters of various orders on the frequency in the range from 1 Hz to 5 Hz.

The analysis of **Figure 7** shows that the selected criterion for ensuring an acceptable delay in measurement results ($\tau_{delay} \leq 0, 5\tau_{fl}$) corresponds to a digital Butterworth filter with the order $p = 4$ and $p = 6$. Filter with $p = 6$, according to **Figure 5**, has a magnitude–frequency response closer to the ideal, which determines the choice of a digital Butterworth filter with an order $p = 6$for estimating small-scale TEC fluctuations. The values of its coefficients (a_k, b_k) calculated using the digital filter design and analysis software Filter Designer [22] are given in **Table 1**, and the equation describing this filter can be written as

$$\Delta N_{Ti} = \sum_{k=0}^{6} b_k TEC_{comb\ i-k} - \sum_{k=1}^{6} a_k \Delta N_{Ti-k}. \tag{23}$$

Figure 8 shows the results of the evaluation of small-scale TEC fluctuations (ΔN_T) in accordance with Eq. (23), expressed in TECU $= 10^{16}\ \text{m}^{-2}$ units. These results were obtained at the North-Caucasus Federal University (Stavropol, Russia) using the advanced capabilities of the GPStation-6 receiver.

Thus, to evaluate (isolate) the small-scale TEC fluctuations (ΔN_T), it is advisable to use a 6th-order Butterworth digital filter, the output samples of which are formed according to the obtained Eq. (23). Digital filter, with values of minimum and maximum transmission frequencies$f_{fl\ \min} = 1$ Hz and$f_{fl\ \max} = 10$ Hz and sampling frequency$f_s = 50$ Hz, provides a magnitude–frequency response close to ideal (**Figure 5**) with an acceptable value ($\tau_{delay} \leq 0, 5\tau_{fl}$) of the delay introduced by it (**Figure 7**). To implement a digital Butterworth filter with the specified parameters its

k	a_k	b_k
0	1	0.076745906902313671
1	−3.4767608600037727	0
2	5.0801848641096203	−0.23023772070694101
3	−4.2310052826910152	0
4	2.2392861745041328	0.23023772070694101
5	−0.69437337677433475	0
6	0.084273573849621822	−0.076745906902313671

Table 1.
Coefficients of the 6th order Butterworth digital filter with a bandwidth from 1 Hz to 10 Hz and a sampling rate of 50 Hz.

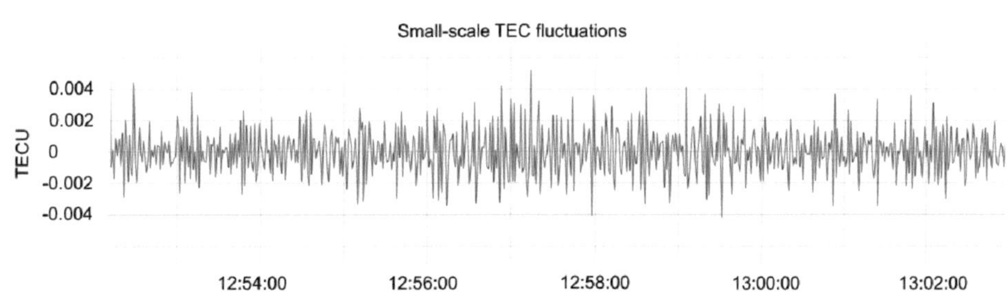

Figure 8.
Small-scale TEC fluctuations at the output of the digital filter.

coefficients (a_k and b_k) are calculated using the digital filter design and analysis tool (Filter Designer) and are shown in **Table 1**.

5. Use of results of GPS monitoring of small-scale fluctuations of the total electronic content of the ionosphere for prediction of the noise immunity of the satellite communication system

Figure 9 shows the simplest case of using the results of GPS monitoring of small-scale TEC fluctuations to predict the noise immunity of the SCS, when the trans-ionospheric radio wave propagation from the GNSS satellite (with carrier frequencies f_1 and f_2) and from the SCS satellite (with carrier frequency f_0) to the GPStation-6 receiver occurs at the same elevation angle ($\alpha \geq 30°$).

In order to estimate the noise immunity of SCS based on small-scale TEC fluctuations (ΔN_T) it is necessary to modify software of GPStation-6 receiver (**Figure 3**) by adding three units (**Figure 9**): RMS ($\sigma_{\Delta N_T}$) calculation unit, scintillation index ($S_4 \sim \sigma_{\Delta N_T}/f_0$) calculation unit, and error probability ($P_{err} \sim S_4$) calculation unit.

Small-scale TEC fluctuations (**Figure 8**) at the output of the digital filter (Eq. (23)) can be characterized by zero mathematical expectation $\overline{\Delta N_T} = 0$ and the RMS

$$\sigma_{\Delta N_T} = \sqrt{\frac{1}{n}\sum_{i=1}^{n}\left(\Delta N_{Ti} - \langle \Delta N_T \rangle\right)^2} = \sqrt{\frac{1}{n}\sum_{i=1}^{n}\Delta N_{Ti}^2}. \quad (24)$$

on the measurement interval $\tau_{RMS} = t_n - t_1 = 1$ s, during which the average (background) value of the TEC ($\overline{N}_T$) can be considered unchanged. The specified measurement interval $\tau_{RMS} = \tau_s n = 1$ s corresponds to $n = 50$ samples with a sampling interval $\tau_s = 0.02$ s (sampling frequency $f_s = 1/\tau_s = 50$ Hz).

The results of the calculation in accordance with Eq. (24) of the RMS ($\sigma_{\Delta N_T}$) of the small-scale TEC fluctuations (ΔN_T) obtained at the output of the digital filter (**Figure 8**) are shown in **Figure 10**.

The analysis of **Figures 8** and **10** shows that at about 12:56:40 local time (UTC + 3) there was a sharp short-term increase in small-scale TEC fluctuations (with RMS up to

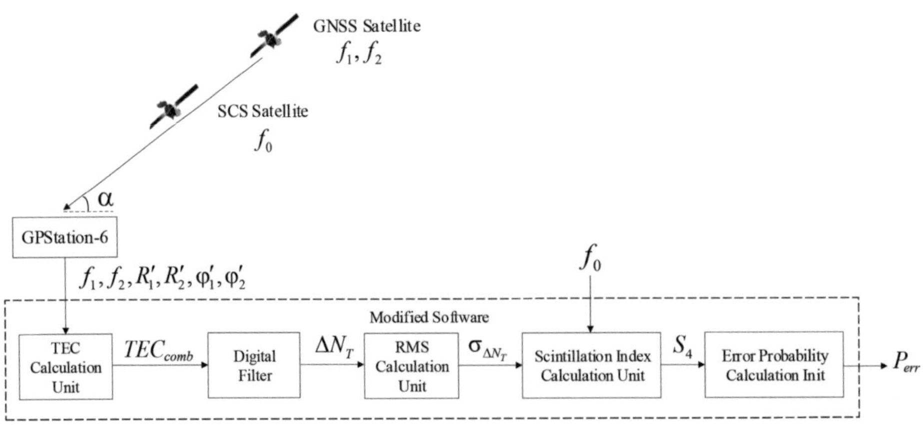

Figure 9.
The structure of the construction of the modified GPStation-6 receiver for the evaluation of the RMS of small-scale TEC fluctuations ($\sigma_{\Delta N_T}$), scintillation index (S_4), and the error probability (P_{err}) when receiving SCS signals.

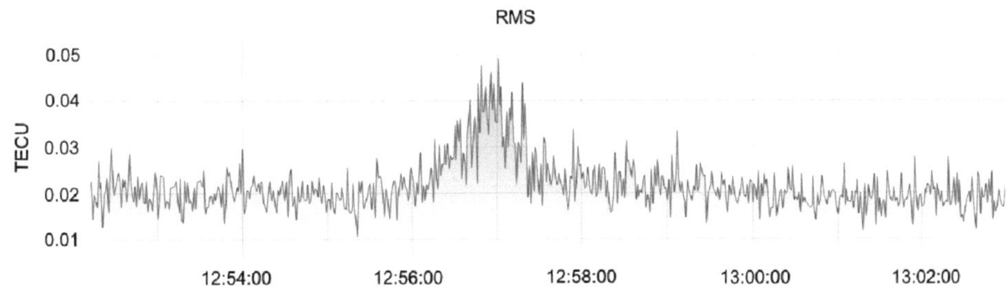

Figure 10.
RMS of the small-scale TEC fluctuations.

$\sigma_{\Delta N_T} \approx 0.05$ TECU). It coincides with the moment when the received GLONASS signals scintillation occurs (**Figure 2**). This generally confirms the dependence of the ionospheric scintillation index ($S_4 \sim \sigma_{\Delta N_T}$) on the RMS ($\sigma_{\Delta N_T}$) of small-scale TEC fluctuations. In the considered interval, it is $\sigma_{\Delta N_T} \approx 0.02$ TECU on average, which is consistent with the known [5, 6, 8, 10] data on the magnitude of relative TEC fluctuations of the mid-latitude ionosphere at typical values of the average TEC $\overline{N}_T \approx 10$ TECU: $\sigma_{\Delta N_T}/\overline{N}_T \sim \sigma_{\Delta N}/\overline{N}_m < 10^{-3}...10^{-2}$. This confirms the reliability of the obtained results of filtration of small-scale TEC fluctuations.

Figure 11 shows calculation results of the scintillation index (S_4) of the received signal in the typical SCS (with carrier frequency $f_0 = 1.6$ GHz) in accordance with Eq. (6) based on the results of the evaluation of the RMS ($\sigma_{\Delta N_T}$) of small-scale TEC fluctuations (**Figure 10**).

The results of the scintillation index calculating in accordance with Eq. (6) shown in **Figure 11** are corresponding to the results shown in **Figure 2** (bottom panel). This allows us to conclude that the dependence (Eq. (6)) is reliable. However, unlike the results of calculating the scintillation index by the built-in GPStation-6 receiver software (bottom panel in **Figure 2**), the results shown in **Figure 11** have two advantages:

1) as initial data for the calculation of the scintillation index S_4 (Eq. (6)), the results of the evaluation of the RMS ($\sigma_{\Delta N_T}$) of small-scale TEC fluctuations are used. It characterizes the cause of the scintillation - small-scale ionospheric irregularities. Multiple reflection of radio waves from objects surrounding the receiving antenna has no effect on the results of calculating the scintillation index;

2) the results of the evaluation of the RMS ($\sigma_{\Delta N_T}$) of small-scale TEC fluctuations and the results of the scintillation index S_4 (Eq. (6)) calculation are formed with a sampling interval $\tau_s = 1$ s. This makes it possible to register an increase in the scintillation index when short-term (units of seconds) ionospheric scintillation occur.

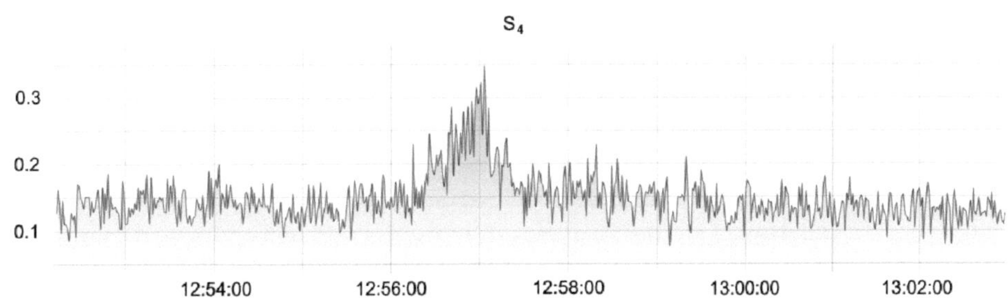

Figure 11.
Scintillation index of received SCS signals with carrier frequency $f_0 = 1.6$ GHz.

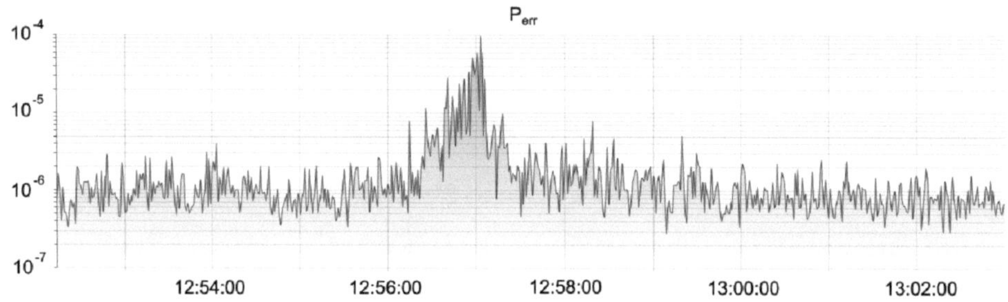

Figure 12.
Probability of error when receiving SCS signals.

The monitoring data of the RMS ($\sigma_{\Delta N_T}$) of small-scale TEC fluctuations allow us to predict changes in the noise immunity of SCS based on Eq. (6) $S_4 \sim \sigma_{\Delta N_T}/f_0$ and Eq. (7) $P_{err} = \psi(S_4, \langle h^2 \rangle)$. The results of calculating the probability of error (P_{err}) when receiving SCS signals with an average signal-to-noise energy ratio $\langle h^2 \rangle = 12$ dB at the receiver input are shown in **Figure 12**.

The analysis of **Figure 12** shows that in the conditions of an undisturbed mid-latitude ionosphere, high noise immunity of the typical SCS is provided. It is characterized by the magnitude of the error probability when receiving signals $P_{err} \approx 10^{-6}$. At the moment of the occurrence of weak ionospheric scintillation value of the error probability briefly increased up to $P_{err} \approx 10^{-5}...10^{-4}$ (by 1–2 order of magnitude). This confirms, that under conditions of natural disturbances of the ionosphere in polar and equatorial regions, the error probability may significantly increase and exceed the permissible value (its traditional value is $P_{err} \leq 10^{-5}$ according to [8]).

The experimental data obtained (**Figures 10–12**) confirm the dependence of the error probability ($P_{err} \sim S_4 \sim \sigma_{\Delta N_T}/f_0$) when receiving elementary symbols of messages in SCS on the RMS ($\sigma_{\Delta N_T}$) of small-scale TEC fluctuations and the choice of carrier frequency (f_0). Ionospheric disturbances accompanied by the formation of small-scale irregularities (and an increase in small-scale TEC fluctuations $\sigma_{\Delta N_T}$) can lead to an increase in the error probability ($P_{err} \sim S_4 \sim \sigma_{\Delta N_T}/f_0$) in SCS using the low-frequency range ($f_0 = 0.2...0.4$ GHz) by more than 2 orders of magnitude (up to $P_{err} = 10^{-3}...10^{-1}$).

6. Conclusion

A method has been developed for GPS monitoring of small-scale fluctuations of the ionospheric TEC and their use for predicting changes in the noise immunity of SCS during ionospheric disturbances. It is based on digital filtering of small-scale fluctuations of the TEC of the ionosphere and evaluation of their RMS.

The software of the GPStation-6 receiver has been modified (**Figure 3**) in the direction of replacing the TEC code measurements with combined (code-phase) measurements, which eliminates noise component of measurement error.

To isolate small-scale TEC fluctuations, it is proposed to use a 6th-order digital Butterworth filter, which at reasonable extremes of the transmission frequency range (1 Hz, 10 Hz) and sampling frequency (50 Hz), provides magnitude–frequency response close to ideal (**Figure 5**) and introduces a delay that does not exceed the permissible level (**Figure 7**). It is described by Eq. (23) and **Table 1**.

The obtained results of the evaluation of small-scale TEC fluctuations at the output of the digital filter (**Figure 8**) allow (in accordance with Eq. (24), Eq. (6), and Eq. (7)) to consistently obtain an estimate of the RMS of these fluctuations (**Figure 10**), the ionospheric scintillation index (**Figure 11**) and the probability of error when receiving elementary symbols of SCS messages (**Figure 12**).

Acknowledgements

The work was carried out with the support of the Russian Science Foundation within the scientific project No. 22-21-00768 (https://rscf.ru/project/22-21-00768) "Methodology for the construction of structural and physical models of trans-ionospheric radio channels and their application to the analysis of satellite radio systems during ionospheric scintillation".

Author details

Vladimir Pashintsev, Mark Peskov*, Dmitry Mikhailov, Mikhail Senokosov and Dmitry Solomonov
North-Caucasus Federal University, Stavropol, Russia

*Address all correspondence to: mvpeskov@hotmail.com

References

[1] Recommendation ITU-R P.531-14. Data on ionospheric propagation data and prediction methods required for the design of satellite services and systems [Internet]. 2020. Available from: https://www.itu.int/dms_pubrec/itu-r/rec/p/R-REC-P.531-14-201908-I!!PDF-R.pdf

[2] Yeh KC, Liu CH. Radio wave scintillations in the ionosphere. Proceedings of the IEEE. 1982;**70**(4): 324-360

[3] Crane RK. Ionospheric scintillation. Proceedings of the IEEE. 1977;**65**(2): 180-199

[4] Aarons J. Global morphology of ionospheric scintillations. Proceedings of the IEEE. 1982;**70**(4):360-378

[5] Kravtsov YA, Feyzulin ZI, Vinogradov AG. Propagation of Radio Waves through the Earth Atmosphere. Moscow: Radio i Svyaz; 1983

[6] Cherenkova E, Chernyshev O. Propagation of Radio Waves. Moscow: Radio i Svyaz; 1984. p. 272

[7] Davis K. Radio Waves in the Ionosphere. Moscow: Radio i Svyaz; 1973

[8] Maslov ON, Pashintsev VP. Models of transionospheric radio channels and noise immunity of space communication systems. Appendix to the Journal Infocommunication Technologies. 2006:4

[9] Pashintsev VP, Peskov MV, Kalmykov IA, Zhuk AP, Toiskin VE. Method for forecasting of interference immunity of low frequency satellite communication systems. AD ALTA: Journal of Interdisciplinary Research. 2020;**10**(1):367-375

[10] Ryzhkina TE, Fedorova LV. Investigation of static and spectral transatmospheric VHF-microwave radio signals. Journal of Radio Electronics. 2001;**2001**:2 Available from: http://jre.cplire.ru/win/feb01/3/text.html

[11] Afraimovich EL, Perevalova NP. GPS monitoring of the earth upper atmosphere. Irkutsk. 2006;**2006**:480

[12] Perevalova NP. Evaluation of the characteristics of a ground-based GPS/GLONASS receiver network designed to monitor ionospheric disturbances of natural and technogenic origin. Solar-terrestrial Physics. 2011;**19**:124-133

[13] Rytov SM, Kravtsov YN, Tatarsky VI. Introduction to Statistical Radiophysics. Vol. 22. Moscow: Nauka; 1978. p. 464

[14] Fremouw EJ, Leadabrand RL, Livingston RC, Cousins MD, Rino CL, Fair BC, et al. Early results from the DNA wideband satellite experiment-complex-signal scintillation. Radio Science. 1978;**13**(1):167-187

[15] Popov VF. Evaluation of the noise immunity of spaced signals reception in the channel with fading according to the Nakagami law and coherent weight addition. Omskiy Nauchny Vestnik. 2012;**3**(113):309-313

[16] Demyanov V, Yasyukevich Y. Space weather: Risk factors for global navigation satellite systems. Solar-Terrestrial Physics. 2021;**7**(2):28-47

[17] Shanmugam S, Jones J, MacAulay A, Van Dierendonck AJ. Evolution to modernized GNSS ionoshperic scintillation and TEC monitoring. In: Proceedings of IEEE/ION Position, Location and Navigation Symposium (PLANS) [Internet]. IEEE; 2012. pp.

265-273. Available from: https://ieee xplore.ieee.org/document/6236891

[18] GPStation-6. GNSS Ionospheric Scintillation and TEC Monitor (GISTM) Receiver User Manual [Internet]. 2012. Available from: https://hexagond ownloads.blob.core.windows.net/public/ Novatel/assets/Documents/Manuals/ om-20000132/om-20000132.pdf

[19] Carrano CS, Groves KM. The GPS segment of the AFRL-SCINDA global network and the challenges of real-time TEC estimation in the equatorial ionosphere. In: Proceedings of the 2006 National Technical Meeting of The Institute of Navigation, Monterey, CA. Manassas, VA: ION; 2006. pp. 1036-1047

[20] OEM6. Firmware Reference Guide [Internet]. 2014. Available from: https:// hexagondownloads.blob.core.windows. net/public/Novatel/assets/Documents/ Manuals/om-20000129/om-20000129. pdf

[21] Bogner RE, Constantinides AG. Introduction to Digital Filtering. London: A Wiley-Interscience Publication; 1975. p. 212

[22] MathWorks. Filter Designer [Internet]. The MathWorks, Inc. 2020. Available from: https://www.math works.com/help/signal/ref/ filterdesigner-app.html

Chapter 3

Nongravitational Accelerations: A Study on the Accelerometers of GRACE and GRACE-FO

Myrto Tzamali and Spiros Pagiatakis

Abstract

The nongravitational accelerations measured onboard spacecraft for the purpose of modeling the Earth's gravitational field but also for the investigation of the upper atmosphere are crucial. This study is focused on two LEO satellites, GRACE and GRACE–FO, which carry an accelerometer that measures the nongravitational accelerations, with the most dominant being the drag and the solar radiation pressure. This study presents the physical models of the nongravitational accelerations presented in the literature for the two missions and investigates how the nongravitational acceleration measurements are affected during different time periods of the solar cycle. In addition, the effect of the penumbra transitions in the three axes of the accelerometers which present as jumps in the measurements are presented. Lastly, the response of the accelerometers is investigated during minor and major geomagnetic storms that appeared during the last two solar cycles, the 24th and the 25th.

Keywords: GRACE, GRACE-FO, nongravitational accelerations, solar radiation pressure, solar cycle

1. Introduction

The Gravity Recovery and Climate Experiment missions (GRACE and GRACE Follow–On) are two missions whose objective is the mapping of the gravitational field of the Earth. GRACE mission launched on 17 March 2002, and for more than 15 years it was monitoring the changes of the gravity field of the Earth. GRACE-FO mission launched in 2018 is dedicated to continuing GRACE's legacy. From the measurements of these two satellite gravity missions, apart from the mapping of the gravitational field, we are able to monitor the distribution and redistribution of the Earth's mass and enhance our knowledge of terrestrial water storage changes, sea level changes, ice sheet and glacier mass balance, and ocean circulations [1, 2].

GRACE and GRACE-FO mission consists of a pair of satellites, named GRACE A and GRACE B for GRACE mission and GRACE C and GRACE D for GRACE-FO mission, at an altitude of $\sim$ 500 km, on the same near-polar orbit, at a 89.5° inclination, with a 220-km distance separation between them. As they cross different areas of the Earth, they sense gravitational changes that are expressed as distance changes between the two spacecraft. These distance changes are measured by a microwave ranging system (K-Band) at a micrometer level of accuracy. This configuration is also referred to as a low-low satellite-to-satellite tracking (SST). Each satellite carries in addition to the microwave ranging system, a Global Positioning System (GPS) [3] that provides the absolute positions of the satellite, a high-precision accelerometer that measures the nongravitational accelerations and attitude sensors that provide measurements of the inertial orientation of the spacecraft [4].

For the accurate determination of the static global gravity field, as well as its time variability, the measurements of nongravitational accelerations are crucial [5]. From the total accelerations calculated from the second derivative of the satellites' position measured by GPS, the nongravitational measurements are subtracted to derive the pure gravitational accelerations. Therefore, both missions carry the SuperSTAR accelerometer.

The electrostatic SuperSTAR accelerometers in both missions were manufactured by ONERA in Paris, France, and measure the nongravitational accelerations acting on each satellite such as the solar radiation pressure (SRP), the drag, the Earth radiation pressure (ERP), and the thermal radiation pressure (TRP). Its principle is that there is a proof mass fixed to center of the mass of the satellite which needs to be centered. In order to always keep the proof mass centered, voltages are applied. This control voltage defines the nongravitational forces acting on the satellite. To ensure that the accelerometer measures purely the nongravitational accelerations, its proof mass is precisely placed in the center of the mass of each satellite and the resolution is 10^{-10} m/s^2 $\sqrt{Hz}$ [6].

GRACE-FO accelerometers are similar to those of GRACE with some improvements to avoid temperature variation effects induced in the measurements [7]. They measure the linear and the angular nongravitational accelerations along three axes in the accelerometer reference frame (ARF) with the Z_{ARF} defined in the along-track direction pointing toward the other satellite, Y_{ARF} nadir-pointing and the X_{ARF} completing the right-handed coordinate system.

The raw acceleration measurements are obtained with a sampling rate of 10 Hz (Level 1A). For the users' convenience, the raw accelerations are filtered with a low-pass filter of 35 mHz corner frequency, time corrected, and under sampled to 1 Hz [8]. The 1-Hz accelerometer data along with most of the processed measurements are given in a dataset called Level 1b in the science reference frame (SRF), whose X_{SRF} is pointing toward the other satellite, Z_{SRF} is nadir-pointing, and Y_{SRF} completes the right-handed triad [1, 4–5]. More details on the satellite frames can be found in [9]. The satellite body-fixed reference frames are shown in **Figure 1**. All the original measurements that are used and will be presented in this study are derived from Level 1b data (1 Hz), of GRACE A (GRACE mission) and GRACE C (GRACE-FO mission), since GRACE A data are of a better quality than GRACE B and GRACE D accelerometer, after 1 month in orbit, presented a high signal-to-noise ratio and since then is underperforming. As a result to this malfunction, a method called transplant method is used to predict the accelerations for GRACE D, using the data of GRACE C. For more information about this method, the readers can be referred to [10, 11].

As mentioned, the nongravitational accelerations acting on the satellites are the SRP, the ERP, the TRP, and the drag, with the most dominant in both missions being the SRP and the drag. Their magnitude is highly correlated with the solar cycle variations due to the solar irradiance which heats the upper atmosphere and alter the drag acting on the satellites [12]. The solar activity varies with an 11-year cycle which is defined from the cycle minima and maxima, corresponding to the increase and

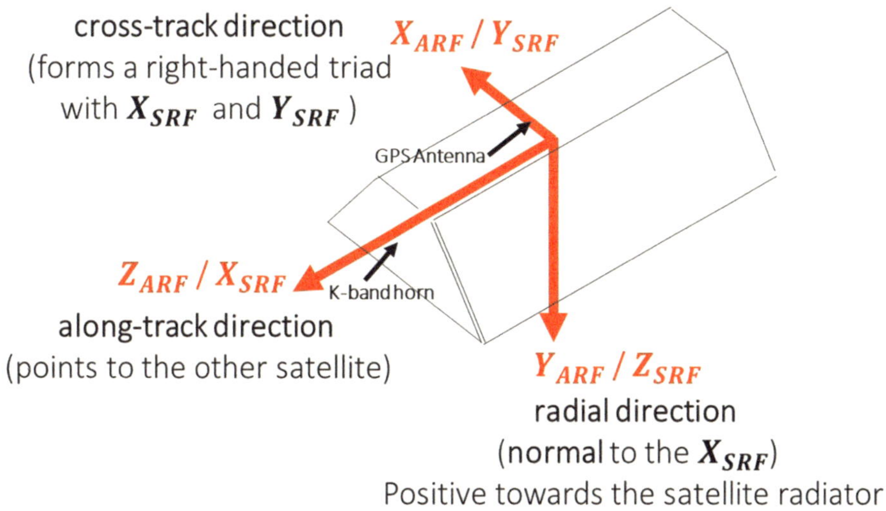

Figure 1.
The satellite body-fixed reference frames in relation to the satellite body.

decrease of the sunspots [13]. GRACE mission was monitoring the Earth from 2002 until 2017, flying during the mid of solar cycle 23 and most of solar cycle 24 while GRACE-FO monitors the Earth since 2018, during the end of solar cycle 24 and the beginning of solar cycle 25. Solar cycle 23 lasted from January 1996 to December 2008 and was magnetically weak [14]. Solar cycle 24 started in December 2008 and ended in December 2019 and had an extended period of a very low sun activity. In fact, solar cycle 24 has been the weakest cycle of the past century, contrastingly to the predictions [15]. The current solar cycle 25 is believed to be similar to the 24 based on observations and predictions [16]. The contribution of the solar cycle effects in the modeling of the nongravitational accelerations will be examined in Section 2.

The modeling of the nongravitational accelerations is crucial for the prediction of a satellite's lifetime, the thermospheric density determination, and the calibration of the accelerometer because the instrument cannot be calibrated on the ground due to strong gravitational signal. For the study of the ionosphere, the accelerometer measurements can be used in the modeling process, to determine the accuracy of the physical models but also to derive the thermospheric densities in the upper atmosphere [17, 18]. Many researchers have proposed different physical models which will be presented in Section 2, with the drag model being the one with the highest uncertainty due to its sensitivity, among others to the solar activity level.

In addition to the physical models that have been proposed for the two missions, an alternative approach to the modeling of the sum of the nongravitational accelerations is presented in this chapter. This new approach is based on the least squares spectral analysis (LSSA), and it can be used for the modeling of the sum of the SRP and the drag acting on the satellite since both forces have a period very close to the period of the satellite. After the modeling of the SRP and the drag, the residual series will be investigated for four severe geomagnetic storms during the operational years of the two missions. The investigation of the residual series of the accelerometer measurements during these storms could enhance our knowledge on how the accelerometer response changes during a geomagnetic storm, the time that the satellite measures the disturbances and their latitudinal signatures. These disturbances on the accelerometer measurements could be used in further studies as an indicator, concurrently with other magnetic and electric field measurements, of how a satellite is affected during a geomagnetic event.

2. Study of the nongravitational accelerations of GRACE and GRACE-FO

In this section, the accelerometer Level 1B data for GRACE A (GRACE mission) and GRACE C (GRACE-FO mission) satellites are presented and compared. Furthermore, the modeling of the sum of the nongravitational accelerations using the least squares spectral analysis is presented, and the residuals series during five severe geomagnetic storms are investigated.

In **Figure 2**, the solar cycle variations are displayed with respect to the sunspot number and the operational years of the two missions. The solar cycle variations are helpful to understand the behavior of the accelerometers but also for the accuracy of the proposed physical models that have been presented for the two missions [19].

2.1 Solar radiation pressure

The SRP accelerations are derived by the nonconservative force acting on the satellite which is produced by the impact of sunlight photons on the surface of the spacecraft. Its modeling is very important because it is one of the two dominant nongravitational forces acting on a satellite. Its magnitude depends on the reflectivity properties of the surface of the satellite, the area of the illuminated surface, the Sun-to-satellite direction, and the mass of the satellite [20]. The SRP effect on actual satellites was calculated for the first time by Parkinson, Jones, and Sapiro in 1960. Ferraz Mello [21] introduced the shadow function ψ which is actually an on-off "switch," and it equals 1 when the satellite is illuminated by the Sun and equals 0

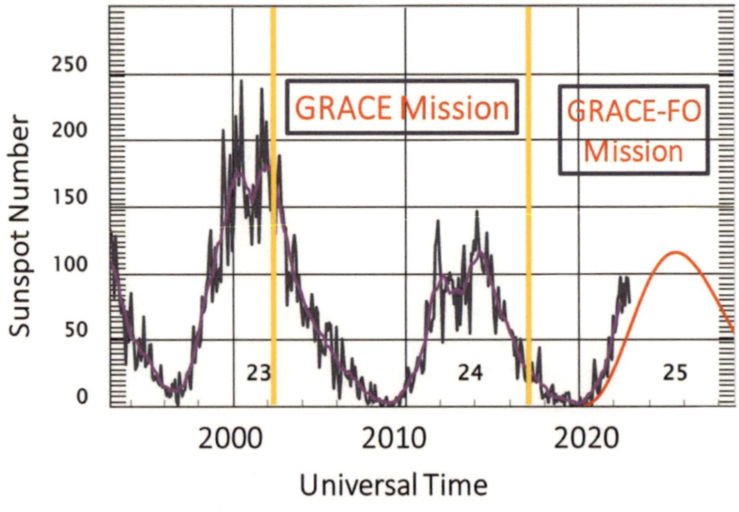

Figure 2.
Sunspot number during the operational years of GRACE and GRACE-FO mission. GRACE mission covered the declining phase of the solar cycle 23 and most of the solar cycle 24, while GRACE-FO mission was launched in the declining phase of the solar cycle 24 and is expected to cover the increasing activity of the solar cycle 25. The plotted data are available from the National Oceanic and Atmospheric Administration (NOAA).

when the satellite is in the shadow. Sehnal [22] addressed the physical and mathematical difficulties in the modeling of the radiation pressure and introduced a special shadow function to describe the effects when the satellite enters and exits the Earth's shadow. In the updated shadow models, the light diffusion due to the refraction, the atmospheric absorption, the effect of ozone, the relative position of the Sun, Earth, and the satellite, and the shape of the conical shadow surface have been considered.

The most commonly used SRP model is the cannonball model which assumes that the satellite's orientation is constant with respect to the Sun. For GRACE mission, a state-of-the-art panel model has been developed that uses the optical properties of the surfaces [23].

2.2 Atmospheric drag

The atmospheric drag force is co-linear to the satellite velocity but in the opposite direction. It arises from the collisions between the satellite and the neutral and charged particles and is the main cause of the deceleration of the satellites and the decrease of their lifetime. The first scientists presented how collisions with neutral and charged particles affect the satellite orbit were Jastrow and Pearse [24]. The modulus of the atmospheric drag is commonly represented with the formula:

$$a_{drag} = -\frac{1}{2}\rho\frac{c_D A}{m}v_{rel}^2\frac{v_{rel}}{|v_{rel}|}, \tag{1}$$

where m is the mass of the satellite, A is the area of the satellite, c_D is the drag coefficient, ρ is the atmospheric density, and v_{rel} is the satellite's velocity relative to the rotating atmosphere. The atmospheric density due to uncertainties caused by different solar activities and the disturbance in the magnetosphere is very difficult to model precisely. Also, the drag coefficient approximation is accurate only when the density and the velocity do not vary much along the satellite's orbit. Lastly, one of the most important variables in the drag determination is the relative velocity which is accurate only during the assumption that the lower part of the atmosphere rotates with the Earth—an assumption that is not truly accurate with the presence of the neutral winds, which can have high velocity and affect dramatically the drag estimation.

For the drag modeling and for the understanding of the solar activity, indices such as $F_{10.7}, A_p, K_p$ are used. One excellent indicator of solar activity is the $F_{10.7}$ radio flux, is measured in solar flux units, and can be measured on a daily basis from the Earth's surface. The A_p index provides a daily average level of the geomagnetic activity from eight daily values [25], while K_p is a 3-hour average quasi-logarithmic index derived from the maximum fluctuations of the horizontal components of Earth's magnetic field. Some of the assumptions made by the scientists to create an atmospheric drag model are the following: the use of the $F_{10.7}$ is a suitable approximation for the atmospheric models and very important in the heating of the upper atmosphere; the lower atmosphere rotates with the Earth; the temperature profiles are based on basic models; the drag coefficient is correct; and all the interpolated variables and indices such as the A_p, K_p do not introduce errors in the models [26].

For the models of the aerodynamic coefficients (dimensionless quantities which characterize the aerodynamic force and moment acting on the satellite), the velocity, the temperature, and the chemical and physical conditions of the satellite surface are taken into consideration. For satellites with complex shapes, the aerodynamic coefficients are calculated by dividing the body of the satellite into segments of simpler shapes and adding them together.

2.3 Earth radiation pressure

Earth radiation pressure contains the Earth's reflected and emitted radiation. The reflected sunlight acts on the satellites and causes ERP accelerations Earth radiation models are calculated based on three main assumptions: The Earth behaves like a Lambertian sphere; the radiation is reflected or emitted; and there is global conservation of energy [27]. The Earth's irradiance (Earth radiation power per unit area) reaches the satellite primarily in the radial direction. The amount of reflected energy that affects the satellite is influenced by local changes in the atmospheric and surface properties of the Earth. The Earth radiation pressure should not be confused with the albedo term, since albedo and emission models are extensively used for the derivation of the ERP. A strict definition of the albedo is that albedo is called the radiative flux that includes the shortwave variations, and it is a measurement of electromagnetic radiation in the visible and near-infrared spectrum [28]. Studies have shown that the Northern and Southern Hemispheres show the same amount of irradiance. Also, there is a seasonal cycle of the surface albedo which attains a maximum in boreal springtime due to the increased reflectivity of the land surfaces been covered by snow between 30°N and 60°N. The maximum of the annual cycles of the irradiance is in March, the second maximum is in October, and the minimum is between June and July [29].

A very detailed study on the modeling of the SRP and ERP for Low Earth Orbit satellites, with an application on GRACE A data, has been published by Vielberg and Kusche [30].

2.4 Review of the physical models proposed for GRACE mission

Many physical models have been proposed for the modeling of the nongravitational accelerations for GRACE mission that can been used for the studies of the upper atmosphere but also for the calibration of the instrument which is impossible to be calibrated before the launch of the satellite due to strong gravitational signal on the ground. The SRP models presented for GRACE are of high precision and almost uncorrelated with the solar activity, since the SRP is unaffected during different solar activity levels contrastingly to the drag models which increase due to the increase of the atmospheric density [31].

One common method for the modeling of the nongravitational accelerations is the satellite acceleration approach [32]. The main idea of this approach is to derive the total accelerations acting on the satellite by double numerical differentiation of the positions estimated from the GPS measurements. This approach, even though it increases the noise of the calculated total accelerations due to the double differentiation, is very accurate as the position of the satellite can be determined with a few centimeter accuracy. Subsequently, modeled gravitational accelerations are subtracted from the total accelerations to estimate the nongravitational accelerations. Bezděk proposed a physical model using the neutral thermospheric density model DTM-2000 and the zonal and seasonal models of the ERPand the shape and physical properties of the satellite. In his work, the modeled nongravitational accelerations are shown during one orbit of GRACE A ($\sim$ 91minutes) on 11 August 2003 (high solar activity), and it is observed that the modeled drag is the force that affects the most the satellite in the along-track direction ($\sim 90\ nm/s^2$ magnitude), the SRP affects the cross-track direction the most ($\sim 30\ nm/s^2$), and the SRP and the albedo are dominant in the radial direction (~ 50 and $\sim 10\ nm/s^2$, respectively) with the drag acceleration being almost 0.

Another high-precision model has been presented for GRACE mission that is used for the calibration of the instrument [19]. In this model, the researchers implemented their model using a tool named XHPS, developed at ZARM institude, and they based their proposed model on a detailed finite element of the satellite. A finite element model is used for the calculation of the thermal radiation pressure. The physical models for nongravitational accelerations have been presented during high (March 2003) and low (February 2009) solar activity. During high solar activity, the drag in the along-track direction is $\sim 100\ nm/s^2$, while the SRP is $\sim 40\ nm/s^2$ and similar to the SRP during low solar activity contranstingly to the drag which is one order of magnitude smaller.

Along the cross-track direction, the drag is ~ 10 and $\sim 1\ nm/s^2$ for high and low solar activity, respectively. The SRP is $\sim 30\ nm/s^2$. In the radial direction, the SRP is $\sim 40\ nm/s^2$, the albedo $\sim 10\ nm/s^2$, and the thermal radiation $\sim 5\ nm/s^2$. There are no significant changes in the radial direction between the high and the low activity.

All the proposed physical models are based on the satellite's geometry and surface and material properties which are given in the GRACE Product Specification Document [33]. The atmospheric drag models are either based on the DTM2013 [34] or the JB2008 [35], and the albedo model is based on CERES [36].

2.5 GRACE A data from 2004 to 2009

As mentioned in Section 1, all the data used in this study are the Level 1B data from GRACE A and GRACE D which are available for both missions at https://podaac-tools.jpl.nasa.gov/drive.

The period 2004–2009 has been chosen to depict the nongravitational accelerations as measured by GRACE-A satellite as they change during the transition from the maximum phase of the solar cycle 23 to its minimum and then to the beginning of cycle 24 (January 2009). In **Figure 3**, the measurements in the three axes of the Science Reference Frame (SRF) are presented.

The dominant nongravitational accelerations in the X_{SRF} are the drag and the SRP. It is clear that the amplitude of the accelerations is highly correlated with the phase of the solar cycle as it is $\sim 10\ nm/s^2$ larger when the sunspot number is ~ 100, and it decreases when the sunspot number approaches zero. This is due to the fact that during high solar activity, the drag acting on the satellite increases. In 2004 and 2005, the measurements show high disturbances compared to the years after 2006 due to the higher solar activity but also due to the presence of severe geomagnetic storms of a $K_p\ index = 8$. There is a strong periodicity of ~ 161 days, which is more visible after 2006 because of the less disturbed signal. This periodicity in the accelerations is because the accelerometer is strongly correlated with the β' angle which is the angle between the satellite's orbital plane and the geocentric position vector of the sun. This angle defines the time that the satellite spends in direct sunlight, and as it is close to $+90°$or $-90°$ the satellite is in a full sun orbit meaning that it does not enter the Earth's shadow.

The dominant accelerations in the Y_{SRF} are the SRP and the thermal radiation pressure (TRP) with the drag and the ERP being close to zero (less than $1\ nm/s^2$). The cross-track axis are the least sensitive axis, but also the noisiest due to thruster activations, especially in the equatorial region and the poles. In this axis instead of the 161-day periodicity due to the β' angle, there is a 90-day periodicity because the

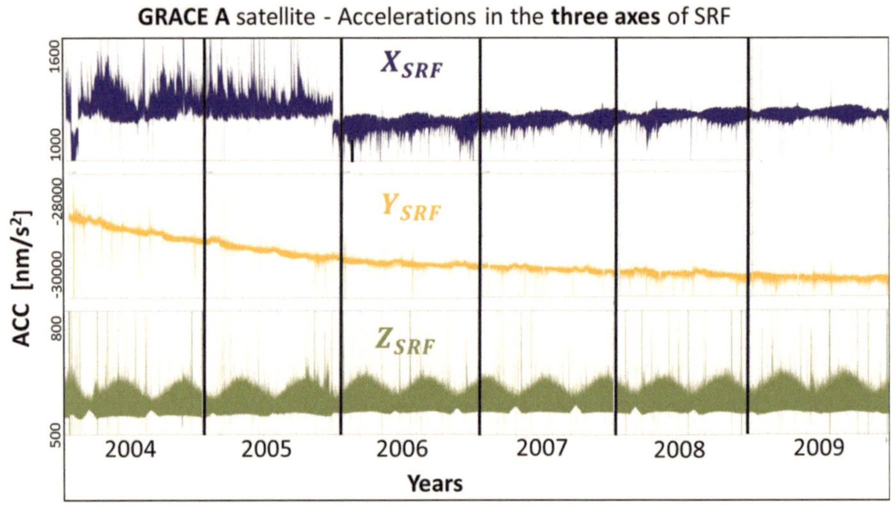

Figure 3.
Total nongravitational accelerations along the three axes of the SRF as measured by GRACE-A in the period 2004–2009. In blue are the accelerations in the along-track direction (X_{SRF}), in yellow are the accelerations in the cross-track direction (Y_{SRF}), and in green are the accelerations in the radial direction (Z_{SRF}).

minimum amplitude of the accelerations is during the full sun orbit and when the sun is along the radial direction, meaning that the accelerations in the cross-track direction are zero when the accelerations in the radial direction are maximum.

The dominant accelerations in the Z_{SRF} are the SRP, the thermal radiation pressure, and the ERP. Similarly, to the other two axes, there is the strong periodicity correlated with the β' angle.

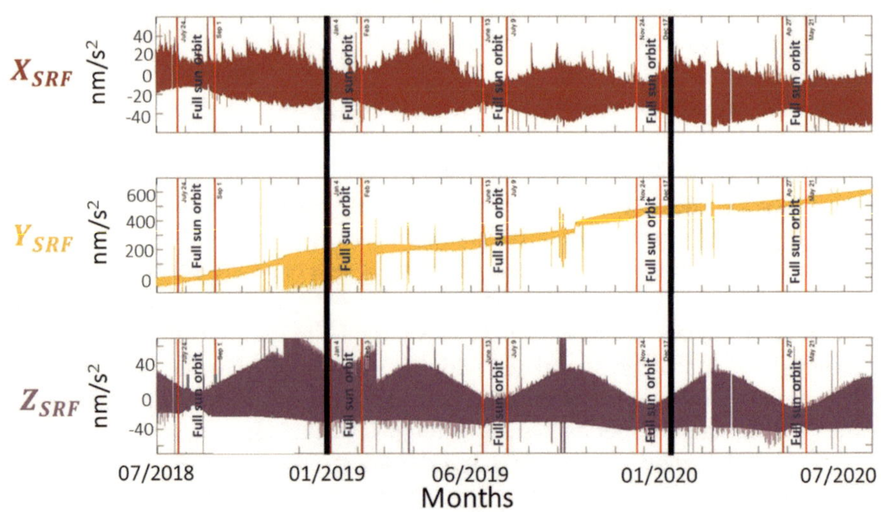

Figure 4.
The nongravitational accelerations as measured by GRACE C from July 2018 to July 2020 in the along-track direction(red), the cross-track direction(yellow), and the radial direction (purple). The periods where the satellite is in a full sun orbit are shown with red lines.

When the satellite is in a full sun orbit, the SRP is the minimum; therefore, the magnitude of the accelerations is the lowest. During the full sun orbit, the satellite does not enter and exit the Earth's shadow; therefore, the characteristic jumps in the measurements called penumbra transitions are not present.

For GRACE-FO mission, GRACE C measurements are shown for the period July 2018 to July 2020. GRACE C data have been selected since the accelerometer of GRACE D is not working properly. In **Figure 4**, the nongravitational accelerations along the three axes are shown. GRACE C enters a full sun orbit every $\sim$ 158 days compared to GRACE A which enters a full sun orbit every $\sim$ 161 days. The performance of the accelerometer in GRACE C is significantly better in the X_{SRF} than GRACE A, but the other two axes present spurious accelerations due to thruster activations.

2.6 Penumbra transitions

All satellites that carry an accelerometer on board show in their measurements perturbative effects that are caused by the penumbra transitions during their passages through the Earth's shadow. Both GRACE and GRACE-FO accelerometer measurements are affected by the penumbra transitions on all three axes, and these measured disturbances are very useful, because they can be used to identify the passage of the satellite into the Earth's shadow and its exit from it. This separation is very important due to the different nongravitational forces acting on the satellite during the sun segment of the orbit and the shadow segment (umbra). The importance of the eclipse transitions and their modeling was recognized by many authors that studied the effects of SRP on the motion of satellites [37–40].

The accurate modeling of penumbra transitions is very challenging because it depends on the different characteristics of the satellite mission studied. Different missions due to different orbit inclination, attitude, and spacecraft shape experience different transitions, and this could cause errors in the SRP models. Therefore, misalignments in the SRP

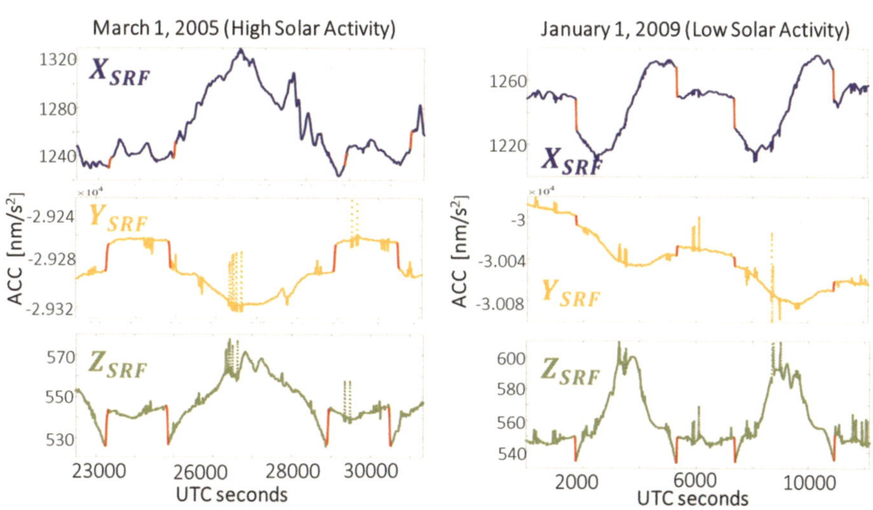

Figure 5.
Accelerometer data of GRACE-A for two orbital revolutions on 1 March 2005(left) and on 1 January 2009 (right). The penumbra transitions appear as jumps in the measurements (indicated in red) when the satellite enters and exits the Earth's shadow.

models and the estimation of the penumbra transitions between the models and the real measurements could introduce error in the calibration process of the accelerometer instrument and consequently in the extraction of gravity field models or thermospheric winds [18, 30, 41, 42]. The appearance of the transitions is correlated with the β' angle, and the existing theoretical models present the largest deviations from the real measurements during periods of β' angle being close to zero. During these periods, the satellite is exposed to direct sunlight for approximately half of its full orbit, and the rest of the time it passes through the Earth's shadow, causing large temperature differences [43].

In **Figure 5**, the accelerometer data for GRACE A are shown. For GRACE, the penumbra transitions in the X_{SRF} are only visible during low solar activity since the drag acting in the along-track direction is smaller than the SRP. In the other two axes, the penumbra transitions are visible throughout all the operational years.

2.7 Modeling the total radiation pressure using least squares spectral analysis (LSSA)

The drag and the radiation pressure have dominant frequencies close to the orbital period (1 cycle per revolution) and to the semi-period (2cpr) of the satellite [4]. This allows for the modeling of the dominant nongravitational accelerations using the LSSA in the frequency domain. It is very important to note here that this analysis is based only on the measurements from the satellite, without the use of any physical model. To the best of our knowledge, this is the first study presented that models the nongravitational accelerations using the spectral characteristics of the measurements. The LSSA is used in this study to analyze both equally and unequally spaced time series contrastingly to Fourier analysis which can be used only when a time series is equally spaced and stationary [44, 45]. The software used for this analysis is in MATLAB code called LSWAVE [46] which can analyze any equally or unequally spaced time series, nonstationary or stationary. This analysis is based on the least square spectrum which provides the best measure of the power contributed by the different frequencies to the variance of the data [47]. In each trial, the least squares spectrum is calculated simultaneously for the amplitudes of all the known constituents that have been used as an input in the software. The inputs in the software are the datum shift, the linear trend, the starting time index of the penumbra transitions, the orbital frequency, and its four harmonics, and they are summarized in **Table 1**. After each trial, the amplitude and the phase of the sin wave of the frequencies are computed. The result after suppressing the five dominant frequencies, the datum shift, the linear trend, and the correction of the jumps in the data due to the penumbra transitions is a residual series that contains shorter drag and ERP variations.

Datum shift (refers to the mean = 0)	
Linear trend	
Starting time of the penumbra transitions as an index	
Frequency 1 (orbital period)	0.1760 *mHz*
Frequency 2 (1st harmonic)	0.3525 *mHz*
Frequency 3 (2nd harmonic)	0.5287 *mHz*
Frequency 4 (3rd harmonic)	0.7004 *mHz*
Frequency 5 (4th harmonic)	0.8810 *mHz*

Table 1.
The known constituents as an input in the LSSA.

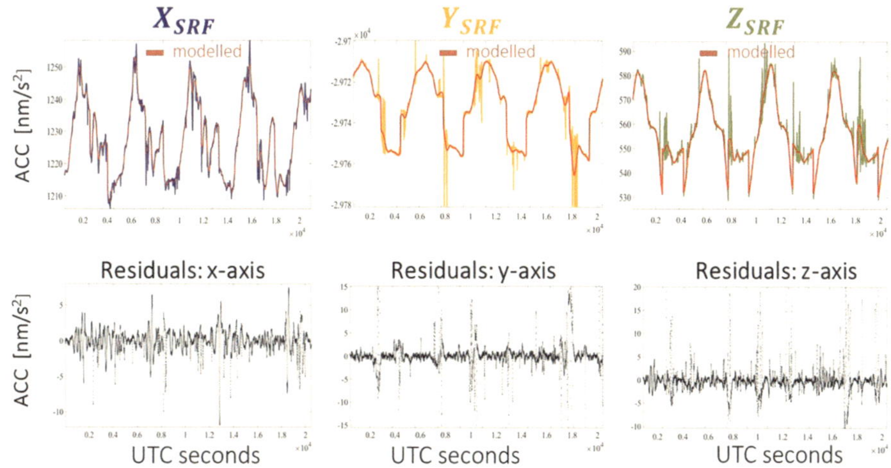

Figure 6.
Top: The original measurements from the accelerometer (Level 1b) on 1 November 2009 for the three axes (blue, yellow, and green) as measured by GRACE-A and the modeled sum of the nongravitational accelerations (red). Bottom: The residual series of the three axes.

Since the modeling is based on the low-frequency components, for shorter drag variations in the along-track direction, the variations in the frequency band (0.176–0.900 mHz] have been added to the model derived from the LSSA. As it is shown in **Figure 6**, the residuals in the X_{SRF} are the largest due to residual drag variations and in the Y_{SRF} are the smallest and close to zero (the spikes in the Y_{SRF} and Z_{SRF} refer to the thruster activations for the attitude correction, measured by the accelerometer).

2.8 Geomagnetic storms as observed by GRACE and GRACE-FO

After the modeling of the total nongravitational accelerations using the LSSA, the scope of this study is to investigate the behavior of the accelerometer on both missions during four severe geomagnetic storms. After the suppression of the dominant frequencies due to the drag and SRP, the residual series contain higher frequency disturbances that show the time that the satellite is affected by the geomagnetic storm, the latitudinal behavior of the disturbances, and some signature signals that could be explained as the response of the instrument due to electromagnetic disturbances. Only the disturbances in the X_{SRF} are presented since it is the most disturbed axis during a geomagnetic storm because of the high drag variations.

Geomagnetic storms are one of the most important components for the study of the upper atmosphere and the space weather effects on Earth. They were discovered about 200 years ago, but data for their study are available the last 50 years due to satellite observations. Their investigation is very important since they can cause damage on the satellites, system failures, and navigational problems [48, 49]. Geomagnetic storms occur due to the ionization of the Earth's atmosphere from the solar flares. The energetic particles interact with the Earth's upper atmosphere and increase the temperature, and as a result the heating increase the drag acting on the satellites. The accelerometer data from both GRACE missions have been used for the investigation and the extraction of the thermospheric densities [50–53]. Densities are used to study the response of the atmosphere to the geomagnetic activity and investigate variations in the interplanetary magnetic field (IMF), high- and low-latitude responses, hemispheric asymmetries, and variations in the solar wind pressure. In this section, the response of the accelerometer in the X_{SRF} will be presented during four severe geomagnetic storms for GRACE and GRACE-FO. Geomagnetic Storms 8–10 November 2004.

2.8.1 Geomagnetic Storms 8-10 November 2004

Two geomagnetic storms occurred during 8 November and 10 November 2004 due to multiple interacting coronal mass ejections (CMEs), which is something unusual during the declining phase of the 23rd solar cycle. The D_{st} reached $-373\ nT$ during 7–8 November and $-289\ nT$ during 9–10 November. During both storms, the K_p index was 9–. The first event starts occurring on 7 November around 1400–1600 UT, which reached lower latitudes at about 2300UT on the same day. The second major storm occurred after 1900 UT on 9 November and reached a peak after 0200 UT on 10 November [54, 55].

The disturbances on the accelerometer started on 7 November at 1442 UT at 70°N with a magnitude of $\sim 28\ nm/s^2$, and the maximum disturbance occurred on 8 November at 0500 UT at 70°N reaching a maximum magnitude of $\sim 300\ nm/s^2$. During the first event, the residual show disturbances at low latitudes at 35°N and at mid-latitudes at 55°S.

Between the two events, the disturbances become smaller but on 9 November 2018 around 1000 UT at 70°N, they start become significantly larger until they reach a maximum peak at 2000 UT of $\sim 270\ nm/s^2$ at 82°N. The second storm lasted more than the first and the disturbances on the accelerometer until 10 November at 2100 UT.

In **Figure 7**, the Level 1B measurements, the modeled nongravitational accelerations, and the residuals, with their histograms, are shown. The two events can be easily distinguished in both the original measurements and the residual series.

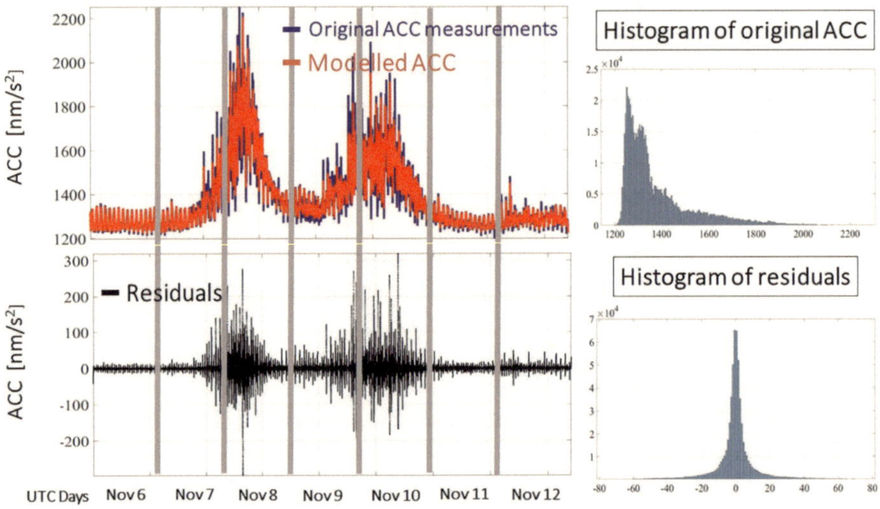

Figure 7.
Top left: The nongravitational accelerations as measured from GRACE A from 6 November to 12 November 2004 (blue) and the modeled accelerations (red). Top right: The histogram of the original measurements of GRACE A. Bottom left: The residual series resulted from the subtraction of the modeled accelerations from the accelerations of GRACE A. Bottom right: The histogram of the residual series.

2.8.2 Geomagnetic storm of 21–22 January 2005

On 20 January 2005, an outstanding solar flare occurred and released a coronal mass ejection creating a geomagnetic storm of a K_p index of 8. The minimum $D_{st} = -105\,nT$ was reached at 0600 UT on 22 January. A strong interplanetary shock created a sudden impulse (SI) at 1712 UT, causing a significant compression of the magnetosphere [56, 57].

In **Figure 8**, the nongravitational accelerations as measured by GRACE A during 20 January to 22 January, the modeled accelerations from the LSSA and the residuals series are displayed along with their histograms. From the measurements of the accelerometer, it is shown that the disturbances in the measurements due to the geomagnetic storm start on 21 January at 1742 UT. The higher disturbances last until 21 January at 2248 UT but are clearly from the residuals that the disturbances are also present on 22 January.

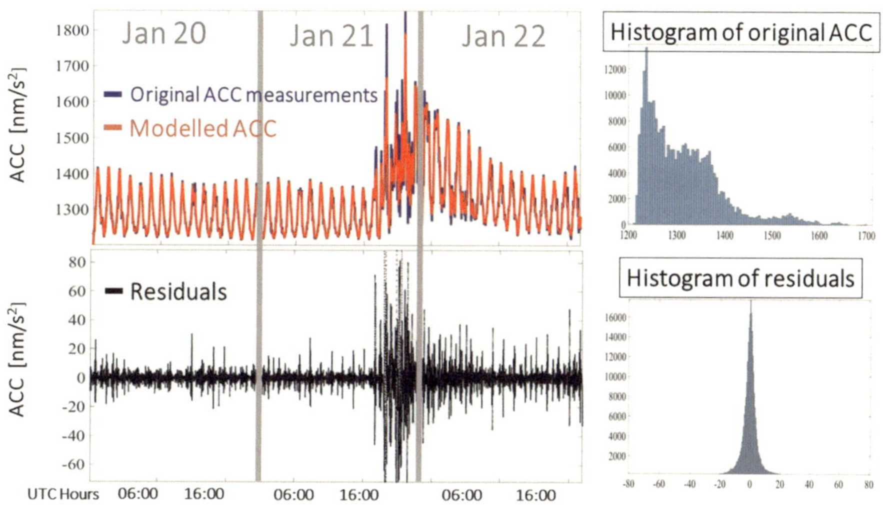

Figure 8.
Top left: Nongravitational accelerations as measured by GRACE A from 20 January 2005 to 22 January 2005 (blue) and the modeled accelerations (red). Top right: The histogram of the original nongravitational accelerations. Bottom left: The residual series after the subtraction of the modeled accelerations from the original measurements of GRACE A.

2.8.3 Geomagnetic storm of 25–26 August 2018

GRACE-FO mission is launched in 2018 at the deep minimum of a rather weak solar cycle 24. On 25–26 August 2018, a G3 class storm occurred with a K_p index of 7. The CME event arrived at Earth on 25 August at 0200 UT and caused a compression on the magnetosphere at 1200 UT. At 1600 UT, a negative direction of B_Z component triggered the main phase of the storm. The minimum $D_{st} = -175\,nT$ was reached at 0600 UT on 26 August [58, 59]. From July 2018 to December 2020, this is the only storm that reached a K_p index of 7. On the residual series, in **Figure 9**, the disturbances started on 25 August at 2100 UTC and reached a peak of $\sim 21\,nm/s^2$ at 0800 UT on 26 August. Disturbances are also shown on 27 August, during the recovery phase of the storm. Comparing the residual series from GRACE-FO and GRACE, it can be observed that residual series on GRACE-F reach a magnitude of $\sim 21\,nm/s^2$,

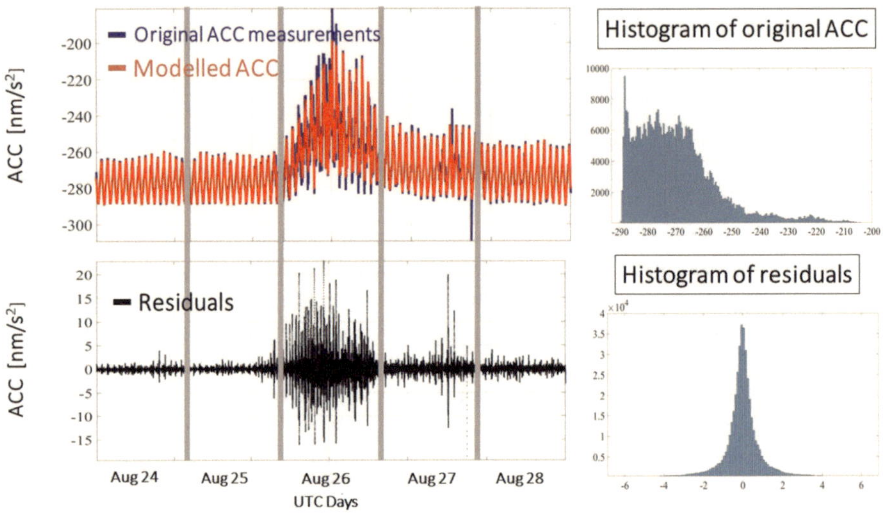

Figure 9.
Top left: Nongravitational accelerations as measured by GRACE C (blue) and the total modeled nongravitational accelerations (red). Top right: The histogram of the original measurements of GRACE C. Bottom left: The residual series after the subtraction of the modeled accelerations from the measured accelerations. Bottom right: The histogram of the residual series.

while for GRACE their magnitude could reach $\sim 300\ nm/s^2$. This can be attributed in the declining phase of the solar cycle 24 since as it was mentioned in Section 2.1, and the magnitude of nongravitational accelerations is highly correlated with the solar cycle variations.

The accelerometer response in both missions during a geomagnetic storm is immediate. Disturbances are visible even after the main phase of the geomagnetic storm due to minor decrease in the $\boldsymbol{B_Z}$ component. The geomagnetic storm in January 2005 was the only case in our analysis where the higher disturbances were present on the Southern Hemisphere instead of the Northern Hemisphere, probably because it was the only storm triggered by a northward direction of the $\boldsymbol{B_Z}$.

3. Conclusions

The purpose of this study was the analysis of the accelerometer measurements of two very important LEO missions, GRACE and GRACE-FO. The correlation between the magnitudes of the measurements with solar activity variations is presented and showed that the measurements are very sensitive to solar activity variations, as expected since they measure, among other components, the atmospheric drag which is highly sensitive to these variations. GRACE-FO measurements are of a better quality in the along-track direction than GRACE mission and of much smaller magnitude, but this could be attributed not only to the instrument performance but also to the operational years that GRACE-FO monitors the Earth (low solar activity).

A literature review of the different models that have been proposed in the modeling of nongravitational accelerations is implemented, showing that the SRP and ERP models present higher accuracy than the atmospheric drag models, which is a rather challenging task due to the complexity of the particle interactions in the upper atmosphere. An alternative method of modeling the sum of the drag and the Radiation

Pressure acting on the satellite using a least squares approach is presented, which is based on the frequency domain. This method can be used for the modeling of the dominant drag and RP accelerations, since the accelerations produced by these two forces show a strong periodicity very close to the orbital period of the satellite. Shorter variations in the frequency band (0.176–0.900 mHz) are attributed to the drag in the along-track axis, since they are very close to the orbital period and its four harmonics. The residual series in the cross-track direction are almost zero, while in the radial direction show structured signals that could be attributed to ERP variations. The along-track direction presents the largest residuals due to remaining drag accelerations in the signal.

The accelerometer response during four geomagnetic storms is investigated. Three severe storms are analyzed during GRACE mission and only one for GRACE-FO, as until December 2020 there was only one severe magnetic event. There is an immediate response in the accelerations even during the initial phase of the storm. The disturbances in the residual series last until the recovery phase, something that could be explained due to the slow recovery of the atmosphere after a severe magnetic event. During the main phase of the storms, disturbances can reach low latitudes in both hemispheres, but the highest are present in the Northern Hemisphere, except from the storm in January 2005. Please note that this study does not claim that accelerometer could measure a direct effect of a geomagnetic event on the satellite. These disturbances could have been caused due to an electrostatic coupling or because the accelerometer detects a mechanical excitation as the satellite flies in a very disturbed and hostile environment which affect its instrumentation.

This study aims to enhance the knowledge of the reader about an instrument very crucial for the studies of the upper atmosphere. Numerous studies have used the accelerometer data for the extraction of the thermospheric densities, and this is beyond the scope of this study. From a preliminary investigation of the residual series of the accelerometer during minor or severe geomagnetic storms, useful time and spatial information of the disturbances are derived and if used together with other electromagnetic field measurements could lead to a better understanding on the numerous interactions that take place in the upper atmosphere.

Acknowledgements

This research was partially supported by the Natural Science and Engineering Research Council (NSERC) of Canada and by a Lassonde School of Engineering Special Research internal grant. The data used by this research can be accessed via podaac.jpl.nasa.gov.

Conflict of interest

The authors declare no conflict of interest.

Author details

Myrto Tzamali* and Spiros Pagiatakis
York University, Toronto, Canada

*Address all correspondence to: myrtotz@yorku.ca

References

[1] Davis E, et al. The GRACE mission: meeting the technical challenges. 1999

[2] Tapley BD, Reigber C. GRACE: A satellite-to-satellite tracking geopotential mapping mission. Bollettino di Geofisica Teorica ed Applicata. 1999;**40**(3–4):291

[3] Dunn C, et al. Instrument of GRACE GPS augments gravity measurements. GPS world. 2003;**14**(2):16-29

[4] Kim J. Simulation study of a low-low satellite-to-satellite tracking mission. Austin: The University of Texas; 2000

[5] Flury J, Bettadpur S, Tapley BD. Precise accelerometry onboard the GRACE gravity field satellite mission. Advances in Space Research. 2008; **42**(8):1414-1423. DOI: 10.1016/j.asr.2008.05.004

[6] Touboul P, Willemenot E, Foulon B, Josselin V. Accelerometers for CHAMP, GRACE and GOCE space missions: Synergy and evolution. Bollettino di Geofisica Teorica ed Applicata. 1999;**40**(3–4):321-327

[7] Christophe B, Boulanger D, Foulon B, Huynh P.-A, Lebat V, Liorzou F, Perrot E. A new generation of ultra-sensitive electrostatic accelerometers for GRACE Follow-on and towards the next generation gravity missions. Acta Astronautica. 2015;**117**:1-7. DOI: 10.1016/j.actaastro.2015.06.021

[8] Wu, S-C, Kruizinga G, Bertiger W. Algorithm theoretical basis document for GRACE level-1B data processing V1. 2. Jet Propulsion Laboratory, California Institute of Technology. 2006

[9] Case K. Gerhard Kruizinga, and Sienchong Wu. GRACE level 1B data product user handbook. JPL Publication D-22027 (2002)

[10] Bandikova T, McCullough C, Kruizinga GL, Save H, Christophe B. GRACE accelerometer data transplant. Advances in Space Research. 2019;**64**(3): 623-644. DOI: 10.1016/j.asr.2019.05.021

[11] Harvey N, McCullough CM, Save H. Modeling GRACE-FO accelerometer data for the version 04 release. Advances in Space Research. 2022;**69**(3): 1393-1407. DOI: 10.1016/j.asr.2021.10.056

[12] Lean J. Variations in the Sun's radiative output. Reviews of Geophysics. 1991;**29**(4):505-535. DOI: 10.1029/91RG01895

[13] Hathaway DH. The Solar Cycle. Living Reviews in Solar Physics. 2015;**12**(4). DOI: 10.1007/lrsp-2015-4

[14] Ataç T, Özgüç A. Overview of the solar activity during solar cycle 23. Solar Physics. 2006;**233**(1):139-153. DOI: 10.1007/s11207-006-1112-3

[15] Nandy D. Progress in solar cycle predictions: sunspot cycles 24–25 in perspective. Solar Physics. 2021;**296**(54). DOI: 10.1007/s11207-021-01797-2

[16] Chowdhury P, Jain R, Ray PC, et al. Prediction of Amplitude and Timing of Solar Cycle 25. Solar Physics. 2021;**296**(69). DOI: 10.1007/s11207-021-01791-8

[17] Sutton EK, Nerem RS, Forbes JM. Density and winds in the thermosphere deduced from accelerometer data. Journal of Spacecraft and Rockets. 2007; **44**(6):1210-1219. DOI: 10.2514/1.28641

[18] Vielberg K, Forootan E, Lück C, Löcher A, Kusche J, Börger K.

Comparison of accelerometer data calibration methods used in thermospheric neutral density estimation. Annales de Geophysique. 2018;**36**(3):761-779. DOI: 10.5194/angeo-36-761-2018

[19] Wöske F, Kato T, Rievers B, List M. GRACE accelerometer calibration by high precision non-gravitational force modeling. Advances in Space Research. 2019;**63**(3):1318-1335. DOI: 10.1016/j.asr.2018.10.025

[20] Montenbruck O, Gill E, Lutze F. Satellite Orbits: Models, Methods, and Applications. ASME. Appl. Mech. Rev. March 2002;**55**(2):B27-B28. DOI: 10.1115/1.1451162

[21] Ferraz Mello S. Analytical study of the Earth's shadowing effects on satellite orbits. Celestial Mechanics. 1972;**5**(1): 80-101. DOI: 10.1007/BF01227825

[22] Sehnal L. Radiation pressure effects in the motion of artificial satellites. Dynamics of Satellites. 1970;**1969**: 262-272. DOI: 10.1007/978-3-642-99966-6_32

[23] Ray V. Advances in Atmospheric Drag Force Modeling for Satellite Orbit Prediction and Density Estimation. Boulder, Colorado, United States: University of Colorado; 2021. ProQuest, Available from: https://ezproxy.library.yorku.ca/login?url=https://www.proquest.com/dissertations-theses/advances-atmospheric-drag-force-modeling/docview/2619259427/se-2

[24] Jastrow R, Pearse C. Atmospheric drag on the satellite. Journal of Geophysical Research. 1957;**62**(3):4-8

[25] Allen J. The Ap index of maximum 24-hour disturbance for storm events: An index description and personal reminiscence by its author. 2004;(no. January): pp. 1–3, 2004.

[26] Vallado DA, Finkleman D. A critical assessment of satellite drag and atmospheric density modeling. Acta Astronautica. 2014;**95**(1):141-165. DOI: 10.1016/j.actaastro.2013.10.005

[27] Rodriguez-Solano CJ, Hugentobler U, Steigenberger P, Lutz S. Impact of earth radiation pressure on GPS position estimates. Journal of Geodesy. 2012;**86**(5):309-317. DOI: 10.1007/s00190-011-0517-4

[28] Knocke PC. Earth radiation pressure effects on satellites (Order No. 8920751). ProQuest Dissertations & Theses Global. 1989:303795978. Available from: https://ezproxy.library.yorku.ca/login?url=https://www.proquest.com/dissertations-theses/earth-radiation-pressure-effects-on-satellites/docview/303795978/se-2

[29] Stephens GL, O'Brien D, Webster PJ, Pilewski P, Kato S, Li JL. The albedo of earth. Reviews of Geophysics. 2015; **53**(1):141-163. DOI: 10.1002/2014RG000449

[30] Vielberg K, Kusche J. Extended forward and inverse modeling of radiation pressure accelerations for LEO satellites. Journal of Geodesy. 2020; **94**(4):1-21. DOI: 10.1007/s00190-020-01368-6

[31] Wang YM, Lean JL, Sheeley NR. Modeling the Sun's magnetic field and irradiance since 1713. The Astrophysical Journal. 2005;**625**(1):522-538. DOI: 10.1086/429689

[32] Bezděk A. Calibration of accelerometers aboard GRACE satellites by comparison with POD-based nongravitational accelerations. Journal of Geodynamics. 2010;**50**(5):410-423. DOI: 10.1016/j.jog.2010.05.001

[33] Lukianova R. *Swarm* field-aligned currents during a severe magnetic storm of September 2017. Annales Geophysicae Discuss. 2020;**38**:191-206. DOI: 10.5194/angeo-38-191-2020

[34] Bruinsma S. The DTM-2013 thermosphere model. Journal of Space Weather and Space Climate. 2015;**5**(A1). DOI: 10.1051/swsc/2015001

[35] Bowman B, Tobiska WK, Marcos F, Huang C, Lin C, Burke W. A New Empirical Thermospheric Density Model JB2008 Using New Solar and Geomagnetic Indices. AIAA 2008-6438. AIAA/AAS Astrodynamics Specialist Conference and Exhibit. 2008. DOI: 10.2514/6.2008-6438

[36] Wielicki BA, et al., Clouds and the Earth's Radiant Energy System (CERES): Algorithm overview. In: IEEE Transactions on Geoscience and Remote Sensing. 1998;**36**(4)1127-1141. DOI: 10.1109/36.701020

[37] Kozai Y. Effects motion of an artificial satellite. SAO Special Report 56. 1961

[38] Fixler SZ. Umbra and penumbra eclipse factors for satellite orbits. AIAA Journal. 1964;**2**(8):1455-1457. DOI: 10.2514/3.2577

[39] Neta B, Vallado D. On satellite umbra/penumbra entry and exit positions. Advances in the Astronautical Sciences. 1997;**95**. PART 2(January): 715-724. DOI: 10.1007/bf03546195

[40] Robertson R, Flury J, Bandikova T, Schilling M, Schilling M. Highly physical penumbra solar radiation pressure modeling with atmospheric effects. Celestial Mechanics and Dynamical Astronomy. 2015;**123**:169-202. DOI: 10.1007/s10569-015-9637-0

[41] Doornbos E, Van Den Ijssel J, Luhr H, Forster M, Koppenwallner G. Neutral density and crosswind determination from arbitrarily oriented multiaxis accelerometers on satellites. Journal of Spacecraft and Rockets. 2010; **47**(4):580-589. DOI: 10.2514/1.48114

[42] Bezděk A, Sebera J, Klokočník J. Validation of swarm accelerometer data by modelled nongravitational forces. Advances in Space Research. 2017; **59**(10):2512-2521. DOI: 10.1016/j.asr.2017.02.037

[43] Behzadpour S, Mayer-Gürr T, Krauss S. GRACE follow-on accelerometer data recovery. Journal of Geophysical Research—Solid Earth. 2021; **2006**:1-17. DOI: 10.1029/2020jb021297

[44] Wells D, Vanicek P, Pagiatakis S. Least—Squares spectral analysis revisited. 1970;**225**(5232):569. DOI: 10.1038/225569a0

[45] Pagiatakis SD. Stochastic significance of peaks in the least-squares spectrum. Journal of Geodesy. 1999; **73**(2):67-78. DOI: 10.1007/s001900050220

[46] Ghaderpour E, Pagiatakis SD. LSWAVE: A MATLAB software for the least-squares wavelet and cross-wavelet analyses. GPS Solutions. 2019;**23**(2):1-8. DOI: 10.1007/s10291-019-0841-3

[47] Barning FJM. The numerical analysis of the light-curve of 12 Lacertae. Bulletin of the Astronomical Institutes of the Netherlands. 1963;**17**:22

[48] Gonzalez WD et al. What is a geomagnetic storm? Journal of Geophysical Research. 1994;**99**(A4): 5771. DOI: 10.1029/93ja02867

[49] Lakhina GS, Tsurutani BT. Geomagnetic storms: historical

perspective to modern view. Geoscience Letters. 2016;**35**. DOI: 10.1186/s40562-016-0037-4

[50] Bruinsma S, Tamagnan D, Biancale R. Atmospheric densities derived from CHAMP/STAR accelerometer observations. Planetary and Space Science. 2004;**52**(4):297-312. DOI: 10.1016/j.pss.2003.11.004

[51] Siemes C, de Teixeira da Encarnação J, Doornbos E, et al. Swarm accelerometer data processing from raw accelerations to thermospheric neutral densities. Earth, Planets and Space. 2016; **68**:92. DOI: 10.1186/s40623-016-0474-5

[52] Bruinsma S, et al. Thermosphere density response to the 20-21 November 2003 solar and geomagnetic storm from CHAMP and GRACE accelerometer data. Journal of Geophysical Research: Space Physics. 2006;**111**(A6)

[53] Krauss S, Behzadpour S, Temmer M, Lhotka C. Exploring thermospheric variations triggered by severe geomagnetic storm on 26 August 2018 using GRACE Follow-On data. Journal of Geophysical Research: Space Physics. 2020;**125**:e2019JA027731. DOI: 10.1029/2019JA027731

[54] Trichtchenko L et al. November 2004 space weather events: Real-time observations and forecasts. Space Weather. 2007;**5**(6):1-17. DOI: 10.1029/2006SW000281

[55] Yermolaev YI et al. Magnetic storm of November, 2004: Solar, interplanetary, and magnetospheric disturbances. Journal of Atmospheric and Solar-Terrestrial Physics. 2008;**70** (2–4):334-341. DOI: 10.1016/j.jastp.2007.08.020

[56] McKenna-Lawlor S, Li L, Dandouras I, Brandt PC, Zheng Y, Barabash S, Bucik R, Kudela K, Balaz J, Strharsky I. Moderate geomagnetic storm (21–22 January 2005) triggered by an outstanding coronal mass ejection viewed via energetic neutral atoms. Journal of Geophysical Research. 2010; **115**:A08213. DOI: 10.1029/2009JA014663

[57] Du AM, Tsurutani BT, Sun W. Anomalous geomagnetic storm of 21-22 January 2005: A storm main phase during northward IMFs. Journal of Geophysical Research. 2008;**113**:A10214. DOI:10.1029/2008JA013284

[58] Piersanti M *et al.*, "From the Sun to 48 the Earth: August 25 , 2018 geomagnetic 49 storm effects." 2020; (no. January): 1-30

[59] Mansilla GA, Zossi MM. Longitudinal variation of the ionospheric response to the 26 August 2018 geomagnetic storm at equatorial/low latitudes. Pure and Applied Geophysics. 2020;**177**(12):5833-5844. DOI: 10.1007/s00024-020-02601-1

Chapter 4

Improving the Accuracy of the Global Navigation Satellite System

Nouf Abd Elmunim

Abstract

The ionosphere influences satellite navigation and communication significantly because it behaves like a dispersive medium. As the density of free electrons increases, so does the influence, and *vice versa*. The concentration of the electron density in the ionospheric layer causes delays that affect the accuracy and reliability of Global Navigation Satellite Systems. Therefore, studying the ionospheric characteristics, variations of the electron density in certain regions, and ionospheric models are crucial when assessing the quality of ionospheric models. These are useful in determining the suitability of the models to improve the accuracy of the global navigation system.

Keywords: GNSS, delay error, ionosphere, models, accuracy of the GNSS

1. Introduction

Global navigation satellite signals are heavily influenced by the environment in space and the Earth's atmosphere because they emanate from satellites at an elevation of around 20,000 km from the Earth to receiving satellites on the surface [1]. The ionosphere extends from approximately 60–1500 km above the surface of the Earth, making it the upper layer of the atmosphere at the edge of space that affects radio communication satellite systems. Within the ionosphere, some solar radiation frequencies are absorbed, resulting in the separation of neutral gas atoms or molecules into free electrons and ions [2], a process known as ionization. The extent to which ionization occurs depends mainly on solar activity and the geomagnetic field. The free electrons and ions from the process interact with microwaves, such as Global Navigation Satellite Systems (GNSS) signals, affecting their propagation velocity and causing a delay of up to several meters in the zenith direction. This process depends on the latitudinal and longitudinal coordinates of the sun-fixed coordinate system, making ionospheric delay important when considering the accuracy of GNSS signals.

GNSS uses one or more frequencies to transmit various signals and navigation data. These signals vary in their availability and dependability. GNSS single-frequency receivers collect satellite radio signals on one frequency and are the most common type of receiver, making them cost-effective but unable to estimate ionospheric delay. Conversely, dual-, triple-, and multi-frequency receivers can collect radio signals from GNSS at a minimum of two frequencies, respectively. Multi-frequency GNSS receivers can estimate ionospheric delays and remove their effects; however, they are used in specialized research areas. Generally, single-frequency is significantly better than dual-frequency in terms of accuracy and cost, a single-frequency receiver may give you decimeters of accuracy within minutes while a dual-frequency receiver may take as long as 20–40 min to reach centimeters of accuracy. Additionally, the single-frequency receiver can outperform the dual-frequency receiver in areas known to have a frequent loss of lock in the signals [3].

2. Global navigation satellite system

GNSS is a satellite constellation revolving around the earth, broadcasting radio signals from within space to GNSS receivers, carrying data for timing and positioning. The system allows receivers to discover their positions within a few meters employing time signals transmitted with satellite radio signals in a linear path. This allows the system to provide global coverage. It comprises three components: the space component (including the constellation of satellites and their movement in space), the control component (which aims to manage the system and owner), and the user component (which consists of the GNSS receivers and user interfaces).

GNSS is classified into two groups: GNSS-1 and GNSS-2. The former is known as first-generation systems that combine the Ground-Based Augmentation (GBA) System or the Satellite-Based Augmentation System (SBAS) with the current satellite navigation system (GLONASS and GPS). Some real-world applications are in the Multi-Functional Satellite Augmentation System (MSAS) used in Japan, the Wide Area Augmentation System (WAAS), a component that is satellite-based in the US, and the European Geostationary Navigation Overlay System (EGNOS) in Europe. On the other hand, GNSS-2, of which the European GALILEO positioning system is an example of a second-generation system that provides a total civilian satellite navigation system. These systems provide the integrity monitoring and accuracy needed for civil navigation. GNSS is the collective consultation of the Global Navigation Satellite System (GLONASS) created by Russia, the Navigation System with Timing and Ranging (NAVSTAR) Global Positioning System (GPS) created in the United States, the Quasi-Zenith Satellite System (QZSS) created in Japan, the Galileo created by the European Union, and the BeiDou developed by China.

GPS is an example of an Intermediate Circular Orbit (ICO) satellite navigation system. The US government's Department of Defense (DoD) developed it in 1973 to assist the military. It is the oldest and most accurate GNSS in the world [4]. The system operates in an L-Band frequency having a radio spectrum between 1 and 2 GHz. The most recent GPS satellite generations employ rubidium clocks with an accuracy of ±5 parts in 1011. More precise ground-based cesium clocks are used to synchronize the rubidium clocks.

3. GNSS/GPS signal structure

Precise cesium or rubidium atomic clocks that oscillate at a fundamental frequency of 10.23 MHz are used to produce the GNSS/GPS satellite signal. The satellites transmit signals on two L-band carrier frequencies, L1 = 1575 MHz and L2 = 1227.60 MHz. The L1 carrier is modulated with two codes: the coarse-acquisition code (C/A) at 1.023 MHz and the precise code (P) at 10.23 MHz. On the other hand, **Figure 1** shows the L2 carrier having only one modulated code, the P-code at 10.23 MHz.

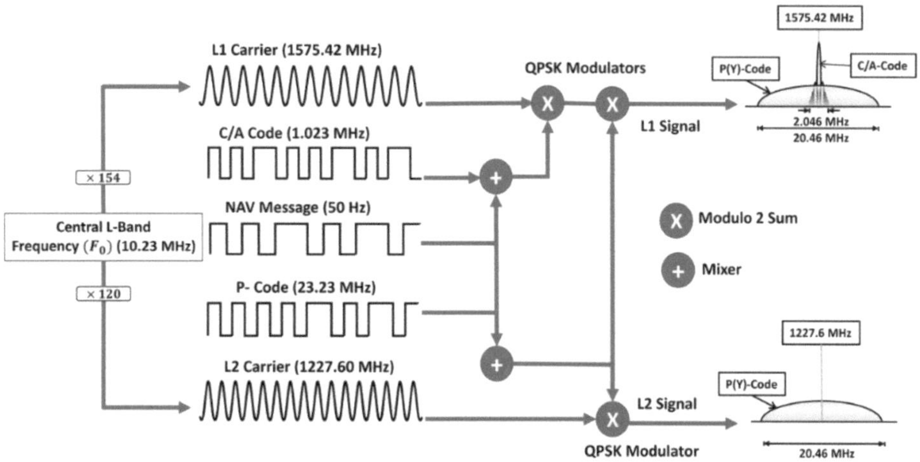

Figure 1.
GPS satellite signals. Source: Khan et al. [5].

The two types of GNSS position determination codes consist of pseudo-random noise (PRN). This noise plays a significant role in the position-determination technique. When modulated on the L1 and L2 carriers, it produces a jam-resistant spread-spectrum signal. Once the anti-spoofing (AS) operation mode is enabled, the P-code may be encrypted into a Y-code. The P-code repeats weekly, beginning at the start of the GPS week, while the C/A code repeats every millisecond. There is also the navigation message, which is modulated with the L1 and L2 carriers at a chip rate of 50 bps. It contains data on ionospheric parameters, perturbations in orbit, satellite orbits, GPS time, satellite clocks, and messages about the system's status. [6]. The quadrature phase-shift keying (QPSK) scheme is used to modulate L1 with P-code, C/A-code, and navigation messages.

To extend the capability of the GNSS, particularly the GPS, it was modernized to give added benefits to the civilian signal. Three new civilian signals, L2C, L5, and L1C, were introduced into the new military signal (M) on the L1 and L2 frequencies. In a dual-frequency receiver, the concurrent use of the L1 C/A (first civilian) and L2C (second civilian) enables ionospheric correction, and enhancing its accuracy in comparison to the military signal. The receivers will increase in complexity in the coming years to support tracking of the new civil code on L2 and the encrypted P on L2 (A-S). L5 has a frequency of 1176.45 MHz and a chipping rate of 10.23 MHz. The L5 code's high chipping rate will provide high-performance ranging capacities as well as better code measurements than L1 C/A code measurements [7]. However, both old and new civil and military codes require improved modulation to better share current frequency allocations with all signals by raising spectral separation and thereby conserving the spectrum. The binary offset carrier (BOC) can then be used to modulate military codes [8].

4. GNSS/GPS observables

There are two fundamental signals in GNSS/GPS observables: carrier phase and pseudorange (code). Pseudoranges are used in high-precision positioning and navigation, where the carrier phase is preferred. Phase observation has the advantage of lower noise but has at least one extra unknown parameter for each satellite, meaning the integer's ambiguity has to be introduced. The difference between the phase of the carrier signal produced by the receiver at the time of signal reception and the phase of the satellite signal produced at the time of signal transmission is known as the ambiguous range or the carrier phase. The initial observation only involves the fractional part of the phase difference. As tracking continues without a cycle slip, the receiver keeps track of the fractional part and the number of cycles. Ambiguity occurs because the measurement does not include the initial number. The carrier phase is typically determined in cycles and can be converted to length units by multiplying it by the wavelength at operational frequencies. Thus, Eq. (1) defines the carrier-phase equation in length units. The pseudorange is the distance between the satellite and the receiver at the start of the code's transmission and reception. It is scaled using a vacuum's nominal speed of light. Timing errors and delays with the GNSS time in the receiver and satellite clocks will differentiate the measured pseudoranges from the LOS, which corresponds to the beginning of transmission and reception. Eq. (2) shows the pseudorange's mathematical expression, and **Table 1** depicts the main characteristics of the carrier phase and the code.

$$\phi_r^s = \rho_r^s + c(\delta t_r - \delta t^s) - I + T + \lambda N_r^s + mp + \varepsilon \tag{1}$$

$$P_r^s = c(t_r - t^s) = c\tau_r^s = \rho_r^s + c(\delta t_r - \delta t^s) + I + T + mp + \varepsilon \tag{2}$$

where ϕ_r^s is the phase measurement in units of length. P_r^s is the pseudorange measured at the receiver. c is the speed of light in a vacuum. $\delta t_r \wedge \delta t^s$ are the satellite and receiver clock errors due to the difference in system time. I is the ionospheric-induced error. T is the tropospheric-induced error. N_r^s is integer ambiguity between the satellite and receiver. λ is the wavelength. mp is the multipath error. ε is the noise or random error, t_r is the reception time of the signal measured by the clock of receivers r, t^s is the transmission time of signal measured by the time frame of satellite s, and τ_r^s is the signal traveling time.

	Carrier	Code
Propagation effects ambiguity	Ambiguous	Unambiguous
Wavelength	20 cm	P-code 30 m C/A-code 293 m
Observation noise	2 mm	P-code 30 cm C/A-code 3 m
Accuracy	High accuracy	Low accuracy

Table 1.
The carrier phases and code characteristics.

Noise or error sources that might be included in the observables are orbital errors, antenna phase center offsets, and receiver noise, which should be considered for measurement accuracy. When the geomagnetic field effect and path bending are ignored, the phase advance is identical to the ionospheric-induced group delay's magnitude. Although they contain ambiguities, the phase observations are two to three orders of magnitude more accurate compared to the pseudorange code. The LOS range between satellite coordinates (X^s, Y^s, Z^s) and receiver coordinates (x_r, y_r, z_r) is provided as:

$$\rho_r^s = \sqrt{(X^s - x_r)^2 + (Y^s - y_r)^2 + (Z^s - z_r)^2} \tag{3}$$

5. Doppler shift

The Doppler shift of a GNSS signal is the carrier phase's derivative of time. It is primarily given by the antenna's relative velocity at the satellite and receiver, and the common offset, which is proportional to the error in the receiver clock's frequency. In order to receive and accept the signals, the GNSS receiver has to estimate the Doppler shift of the signals.

6. GNSS observable errors

Errors and noise caused by the propagation of a signal through atmospheric layers and noise measurements have an impact on phase and code measurements. These errors are classified into three types: satellite-based errors (including satellite clock biases and ephemeris), receiver-based errors (including receiver clock errors, inter-channel biases, and receiver noise), and propagation errors (including tropospheric and ionospheric delays, multipath signals, and other interferences). The main GNSS observable errors are described as follows:

6.1 The satellite-based error

The clock's errors are caused by time differences between the satellite clock and the real GNSS. The transmitted navigation message from the satellite includes some corrections for clock drifts, resulting in positional errors. Satellite clock errors are typically less than 1 msec, which equates to a 300 km error in pseudorange that can be removed by the difference between two receivers based on their respective satellites. Errors in the receiver clocks are significantly bigger than those in satellite atomic clocks because they use inexpensive crystal clocks with reduced precision. Depending on the firmware of the receiver, the size of the receiver clock error will range between 200 ns and some milliseconds. The clock's errors can be eliminated with the difference between the same receiver of two satellites [9]. It is possible to model this using the navigation message carrying a reference time and containing the polynomial coefficients transmitted.

The errors in satellite ephemeris are caused by inaccurate satellite position predictions that are transmitted in a navigation message to users. The satellite's position is dynamic and depends on the solar pressure and gravitational field. The calculation of position by a ground master control station is susceptible to error because of the monitoring station's clock drifts and delays. Errors in estimating the position of the satellite result in pseudorange errors, which must be corrected at the level of the user.

6.2 The receiver-based error

A receiver clock's error occurs because non-precise clocks are used, causing an offset between it and the GNSS reference time. These errors are regarded as unknowns in the pseudorange calculations. However, the double difference equation can be used to remove them, as shown in the following section [6].

The interchange bias is the correlated error that is caused by the internal GNSS receiver processes. It cannot be measured with live data because large errors in the GNSS measurements make the bias impossible to measure because it is much smaller.

The only way it can be measured is to zero out all other channels of error. This can be done using a GNSS simulator [10]. In addition, the biases in GNSS positioning occur because of physical limitations and inadequacies in the hardware of the GNSS. They are caused by little delays between actions that should be run concurrently when signals are transmitted *via* satellite or acquired using a receiver [11]. Furthermore, receiver noise, also known as random measurement noise, influence carrier-phase measurements and code delay. This noise is unrelated to the signal. Noise is produced by the cables, amplifiers, antennae, and the receiver itself. The random measurement noise includes interference from other GNSS signals. Even though the error contributes only a minor portion of overall positioning, it has been improved with modern GNSS receiver technology.

6.3 Propagation error

Multipath occurs when the satellite or receiver gets multiple signal reflections because of the numerous paths the signal takes to reach its destination. The multipath effect depends on the satellite's geometry, the reflecting surfaces surrounding the receiving antenna, and the antenna's position. Signals received from higher-altitude satellites are less susceptible to the multipath effect than those received from low-angle ones. Multipath errors affect the pseudorange measurement much more than they do the carrier phase because of frequency dependency. To reduce multipath errors, a site far from any reflective surfaces (trees, buildings, mounting, and cars) should be selected. Multipath errors, on the other hand, can reach 1 m and can be removed with antennas that use signal polarization to their advantage. The troposphere is the lowest region in the atmosphere where all weather phenomena take place. The GNSS signals are affected by tropospheric attenuation. The tropospheric layer triggers a delay in carrier and code observations that the dual-frequency receiver cannot remove. However, the tropospheric delay can be modeled successfully. Tropospheric models rely on empirical models that take temperature, relative humidity, mapping function, and pressure values into account [12].

The ionosphere is the uppermost layer of the atmosphere, and it contains a large amount of free ions and electrons. GNSS signals are refracted and dispersed nonlinearly in the ionosphere by free electrons. The ionosphere's dispersive nature causes the delays to become dependent on frequency. The number of free electrons passing through the path of the signal ray is proportional to the total ionospheric delay, which varies depending on the time of day, year, solar cycle, and geographic location on and above the Earth's surface [13]. For the GNSS signal frequency receiver, the Klobuchar model can only remove about 50 to 60% of the ionospheric errors. Additionally, users of single-frequency GNSS can use empirical ionospheric models or the regional network *via* communication links using real-time correction. Ionospheric errors on signals received from GNSS satellite signals play the greatest role in positioning errors when using single-frequency GNSS receivers. These phenomena become obvious during severe solar activity, such as geomagnetic storms. For any radiometric space method, like those using the GNSS system, it is imperative to give an account of the propagation delay that emanates from the neutral atmosphere. As the ionosphere is the primary error source in GNSS single-frequency measurements, understanding its variations and properties, as well as identifying accurate ionospheric delay approaches, is critical to improving its accuracy. However, ionospheric errors can be partially eliminated using the linear combination of dual or multiple frequencies [14].

7. Ionospheric total electron content

The total electron content (TEC) is defined by the integral of the electron density in a TEC unit (TECU), where one TECU is equal to 1016 electrons per square meter along the signal transmission path to the receiver. The TEC significantly affects electromagnetic waves that propagate through the ionosphere. TEC is measured using dual-frequency GNSS receivers, which is a popular method for investigating the dynamics of the ionosphere [15]. As a result, the study of the TEC gained a new impetus with GPS-based ground and navigation positioning that makes use of trans-ionospheric communication. As the TEC depends significantly on the state of the ionosphere, it has a large effect on GNSS-based communications. The ionospheric delay varies substantially with the geographical location (equatorial, low-latitude, mid-latitude, and high-latitude regions). However, due to the unique geometry of the magnetic field, the ionospheric delay is obvious in low-latitude and equatorial regions. In such regions, the plasma density in the ionosphere shows significant variations. These variations are influenced by the solar cycle, geomagnetical activities, time of day, seasons, and geographical location [16].

8. Ionospheric regions

The ionosphere is classified into three major regions: the equatorial/low-latitude, mid-latitude, and high-latitude regions. In high-latitude regions, the electron density is significantly lower than in low-latitude regions because solar radiation hits the atmosphere more aggressively there. In mid-latitude regions, ionization occurs due to solar photon radiation, with the electron density not being subject to any particle radiation. In equatorial or low-latitude regions, there is a high electron density concentration because of the high radiation from the sun coupled with the electric and magnetic fields of the earth. Hence, the results of the accumulated electrons in this region influence radio wave propagation, which significantly affects satellite communication systems.

9. Major ionospheric variations

The ionospheric variability is influenced by solar radiation because of photoionization from the sun. Photoionization is caused by the increase in electron density in the ionosphere. The TEC is affected by the activity of the sun, reaching its maximum value when there is high solar activity and decreasing during low solar activity. The sunspot number is one of the indicators commonly employed to determine solar activity. Additionally, the electron density is expected to be higher than that in other locations because of the change in latitude away from the equator and the oblique angle at which solar radiation arrives in the atmosphere. This results in the reduction of ionospheric densities. Geographical locations with low zenith angles are exposed to more solar radiation, hence the higher electron densities. Moreover, maximum electron density occurs during the day and becomes lower at night since there is no photoionization because of the absence of solar radiation. In general, when photoionization stops, recombination significantly reduces the electron density at night, but some free electrons remain until dawn. This variation can appear clearly in equatorial regions. Emunim et al. [16] investigated the daily variation of ionospheric delay at different stations in equatorial regions (UUMK with geographic coordinates at 4.62° N–103.21°E, MUKH with geographic coordinates at 6.46°N to 100.50°E, and TGPG with geographic coordinates at 1.36°N–104.10°E) during a typically quiet day ($Kp \leq 1$) on 7 March 2013. As shown in **Figure 2**, the hourly median local time (Malaysia local time = universal time + eight hours) was charted with the Vertical TEC (VTEC) represented on the horizontal and vertical axes, respectively. The daily pattern of ionospheric delay was observed as a single peak. The general ionospheric delay over all the stations showed that the smallest delay was observed at sunrise, from 4:00 to 6:00 LT. From 13:00 to 17:00 LT, it increased gradually to a maximum. Variations in the solar EUV caused a steady increase in delay, where increased EUV led to increased ionization, which in turn increased the delay and *vice versa*.

Additionally, looking at seasonal variations, during the equinox months, the electron densities increased, because the sun was directly overhead at the equator during solar maximum and minimum activities, leading to an increase in electron density compared to other seasons. Direct sunlight in equatorial regions in equinoctial months causes the most ionization in the ionosphere. The summer had the shortest ionospheric delay, which could be due to uneven heating of the two hemispheres caused by transporting neutral substances from the summer to the winter hemisphere (hot to cold). This reduces the rate of recombination during the winter in comparison to the summer, resulting in a significant rise in the concentration of electrons during the winter. Another possible cause of this phenomenon is a shift in the path of the neutral wind. As it blows in reverse to the plasma diffusion process that begins at the equator, the meridional component of the neutral wind blows from the summer to the winter hemisphere, lowering the ionization crest value in the summer solstice. This results in the equinox meridional winds originating at the equator blowing toward the poles, causing the ionization crest value to be high [17]. An example of the seasonal variation in equatorial regions is discussed in Ref. [18], where Langkawi station was used at geographic coordinates 6.19°N, 99.51°E in 2014. The variabilities of the ionospheric delay (VTEC) during winter (January, February, November, and December), summer (May, June, July, and August), and equinox (March, April, September, and

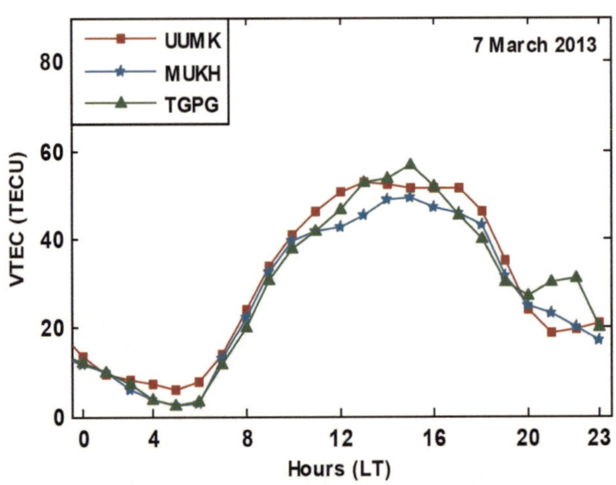

Figure 2.
The diurnal hourly variation of the ionospheric delay. Source Elmunim et al. [16].

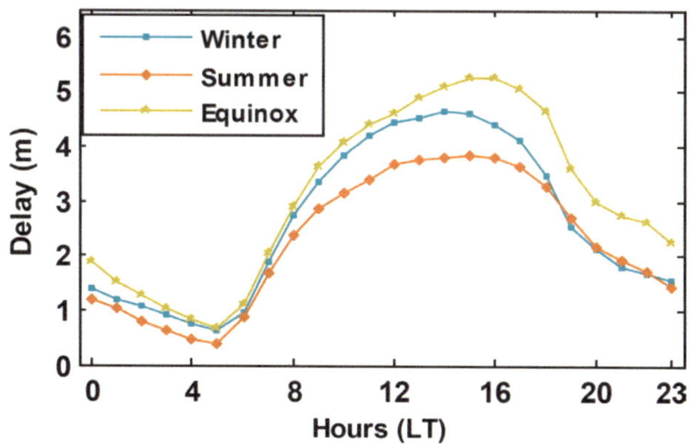

Figure 3.
The seasonal variation of the ionospheric delay.

October) are described in **Figure 3**, where the maximum variation in all the seasons was observed from 9:00 to 19:00 LT. The peak was obtained during the equinox, followed by winter, and finally summer.

Furthermore, during geomagnetic storms, the ionospheric delay affects GNSS signals because of energy input into the polar ionosphere, changing a few of the thermosphere's parameters, such as composition, temperature, and circulation. This change in composition influences the concentration of electrons in the F2 region directly, followed by the TEC as the heated gas is spread to lower latitudes through circulation [19]. During severe geomagnetic storms, currents in the dusk-to-dawn direction that are associated with inner magnetospheric electric fields are no longer able to shield the equatorial and mid-latitudes from the electric fields of high latitude. This causes electric fields to penetrate the ionosphere from high latitudes to the equatorial and middle ionosphere in an instant. So, the transportation of the particles and the instantaneous penetration of the high-latitude electric field toward the equator at high velocities can increase the ionospheric delay and the TEC value [20].

However, this complex nature of the equatorial ionosphere has been of great concern due to the large errors that are peculiar to the GNSS applications [20]. The delay errors are attributed to the free electrons in the signal path. The variations in the value of the TEC need to be considered in the equatorial region so that the accuracy of the GNSS can be improved. Studying the variations in the TEC at different geographical locations under various geophysical conditions like geomagnetic storms is also crucial. Therefore, it is essential to develop a suitable ionospheric model to minimize GPS positioning errors.

Research has been done on different low-latitude and equatorial regions of the world using GNSS. Researchers have documented concerns in the analysis of GPS-TEC variations in a particular ionospheric region at a specific time. For instance, as discovered by Bagiya et al. [21], semiannual variations with maximum values were recorded between March and April and again between September and October. Furthermore, they reported that seasonal variations in the TEC at night show a semiannual periodicity, with high background levels during the equinoctial months and low background levels during the solstices. Tariku [22] discovered that the lowest and highest monthly values of ionospheric delay were seen during the June solstice and the equinoctial months, respectively, from 2009 to 2011 over the Kenyan region.

Shimeis et al. [23] studied the temporary variations of the ionospheric delay at some stations in Africa from 2002 to 2012. They observed that the ionospheric delay's daily variation gave multiple maxima in the summer equinox between 2006 and 2012 and during the equinox months from 2005 to 2008. Based on their report, increasing nighttime VTEC and winter irregularities were observed during the study period. They also reported that, during the time of a deep solar minimum (2006–2009), during a quiet period of geomagnetic activity in the equatorial latitudes, a similar diurnal variation of the ionospheric delay was observed over Alexandria (Egypt). Akala et al. [23] compared equatorial GPS-TEC data from an American and an African station during the ascending and descending phases of the solar cycle 24. The study revealed that the minimum and maximum seasonal values of the ionospheric delay were reported during the solstices and the March equinox, in the solar minimum phases of both stations. Meanwhile, during the ascending phase of solar activity, the June solstice had the lowest values and the December solstice had the highest values. Huy et al. [24] investigated the time variations of the ionospheric delay in Southeast Asian equatorial ionization anomalies from 2006 to 2011. Their findings also revealed that crest amplitude ionization was significantly reduced in a few months during the winter and summer during the deep solar minimum from 2008 to 2009. Based on their report, the two crests substantially move toward the equator in the winter, unlike in other seasons. It is also possible for the two crests to appear earlier in the winter and later in the summer. Furthermore, the study [25] revealed that the daytime seasonal variations of the TEC showed a semiannual periodicity and a minimum value during the winter in the Indian regions. Moreover, at a low-latitude Singaporean station [26], the GPS-TEC values decreased between 15:00 and 16:30 local time (LT) and started to increase between 16:30 and 21:30 LT in 2011. They also reported that the highest TEC values were recorded during the equinoctial months of 2010. Interestingly, similar findings were obtained using various satellites and techniques [16, 19, 27–30].

10. Ionospheric correction models

There are two types of ionospheric correction models: empirical models and time series models. The previous uses long-term data and represents typical patterns of variations, whereas the following is fitted with mathematical functions and uses actual measured ionospheric delay for a specific area over a period.

10.1 Empirical models

The empirical models can provide three-dimensional ionospheric electron profiles and the TEC along the ray's path of GNSS electromagnetic waves. However, because they are derived from monthly median parameters, they only describe the long-term averaged condition of the ionosphere [31]. On the other hand, the grid ionospheric vertical delays broadcasted by the Satellite-Based Augmentation System (SBAS) can provide ionospheric correction with higher accuracy [32]. The broadcast ionospheric models have a few parameters that are easy and widely used for ionospheric delay correction for single-frequency GNSS users, such as the Klobuchar model [33], the developed Klobuchar model [34], the NeQuick model [35], the NTCM model and its modification MNTCM-BC [36, 37], or the BDS broadcast model and its improvements [38].

The GIM presumes that the ionosphere is made up of a thin spherical layer located 450 km from the Earth's surface [39]. It gives VTEC at specific grid points around the world. When reconstructed from measurements of the predicted and actual datasets, it is possible to use them for real-time and post-processed applications, respectively. However, currently, its accuracy is slightly worse than that of the 1-day-predicted GIM, the International GNSS Service (IGS), and the latest GIM (IGSG) [40], because of the limitations of using two-dimensional ionospheric models [41].

The Broadcast Klobuchar ionospheric model was created at the Air Force Geophysics Laboratory by John Klobuchar. The model rectifies ionospheric delays by broadcasting ephemerides, with the benefits of a simple structure and convenient calculations [3]. Having set the parameters and reflecting on the varying characteristics of the ionosphere, the model takes into account the daily periodic and amplitude ionospheric variations to ensure the reliability of large-scale ionospheric forecasts. It assumes an ideal smooth behavior of the ionosphere; thus, daily substantial alterations could not be modeled properly. However, the computational memory and capacity of the model are limited. Hence, the model can only remove around 50–60% of the ionospheric delay errors [42]. As a result, even for absolute positioning, its accuracy is only marginal, especially as solar activity increases [43]. The model has proven to be useful in mid-latitude regions; however, its accuracy in the correction of ionospheric delay is limited.

The International Reference Ionosphere (IRI) model, which was posited for application internationally by the Committee on Space Research (COSPAR) and the International Union of Radio Science, is one of the most widely used models (URSI). Its primary function is the specification of ionospheric parameters [44]. The IRI defines averages once a month of electron density, ion temperature, electron temperature, ion composition, and ion drift at heights ranging between 60 km and 1500 km for given times, dates, and locations. The development of this model is still in progress. The IRI's capacity for determining and predicting ionospheric behavior is continuously updated by the scientific community. The current version of the IRI is the IRI-2016 model, which was released with several modifications [45]. The IRI-2016 model has three topside electron density options, namely IRI-2001, IRI01-corr, and NeQuick. Various topside electron density options are expressed by various mathematical functions. The electron density profile with the NeQuick and IRI01-corr topside options decreases with altitude more swiftly than that with the IRI-2001 topside option [46]. Many researchers worldwide have compared these different topside electron density options to identify the best option and tested the latest IRI version in certain regions [47–49]. The performance of the latest IRI-2016 model using three topside options was compared to the IRI-2012 model at different stations on the equator. They observed that the IRI-2016 model showed some improvements in comparison with the IRI-2012 model and the IRI01-corr; however, Elmunim et al. [16] displayed a minimum error compared to the IRI-2001 and NeQuick topside options. **Figure 4** depicts the hourly variation on a quiet day (7 March 2013) in an equatorial region over the UUMK, MUKH, and TGPG stations. The first row shows the variation by the hour at the UUMK, MUKH, and TGPG stations. The vertical axis represents the VTEC values (TECU), while the horizontal axis displays the hours. The topside options of the IRI-2016 are represented by a solid line, while the topside options of the IRI-2012 are represented by dotted lines, showing the IRI-2001, which is in green, IRI01-corr in black, and the NeQuick in pink. When compared to IRI-2016 and IRI-2012, the IRI01-corr and NeQuick offered better predictive ability during the

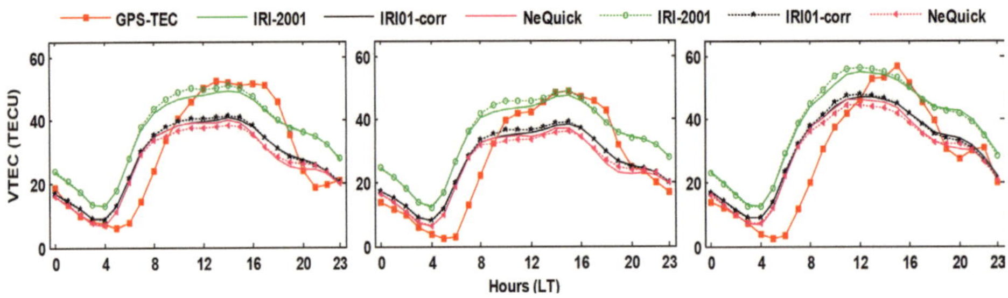

Figure 4.
The diurnal hourly variation of IRI-2016 and IRI-2012 topside options. IRI-2016 (IRI-2001, IRI01-corr, and NeQuick) are plotted in solid lines, while IRI-2012 (IRI-2001, IRI01-corr, and NeQuick) are plotted in dotted lines.

day, except for the afternoon, when the IRI-2001 model provided a better result prediction. In comparison to the IRI-2001 and NeQuick options, the IRI-2016 and IRI-2012 models of the IRI01-corr option produced smaller percentage deviation values. The IRI-2016 outperformed the IRI-2012 in the IRI01-corr and NeQuick options in the majority of the daily disturbed and quiet periods.

10.2 Time series models

The time series statistical models have been developed to estimate and correct the ionospheric delay and for ionospheric forecasting. A time series is a collection of observations with equal time spacing that are measured sequentially. The observation of GNSS data, in this case, the double difference in the ionosphere between a reference and a roving satellite on a specific baseline, is a strong example of such a series. Time series are analyzed as tools for understanding the primary structure and function that created the data. Based on the understanding of these mechanisms, a time series can be mathematically modeled to forecast future observations. Therefore, some researchers have successfully developed a series of high-precision ionospheric prediction models, such as the time series models of Refs. [50–53]. Among these models, the time series model has several advantages over other modeling techniques. It uses less sampling data, includes a complete modeling theory, has a relatively simple calculation process, has good extensionality, and delivers high-precision short-term ionospheric predictions [53].

In addition, the Autoregressive-Integrated Moving Average (ARIMA) method is a time series technique that was created by Box and Jenkins in 1976. This technique allows the production of a set of weighted coefficients that describe the ionosphere's behavior or rate of change during the sample period. These coefficients can then be used to forecast future observations. Serval researchers used the model to correct the ionospheric delay in different equatorial regions [30, 54].

Additionally, the Holt-Winter approach is a statistical time series approach used to forecast future data values to generate repeated trends and a seasonal pattern time series [55]. The exponential smoothing function method is applied to reduce variations in the time series data to provide a clearer viewpoint [56]. The model has additive and multiplicative options. The multiplicative options showed better results with high accuracy compared to the additive option [57]. The model was developed toward forecasting ionospheric delays and thus enhances positional accuracy by up to 95% in equatorial regions [19].

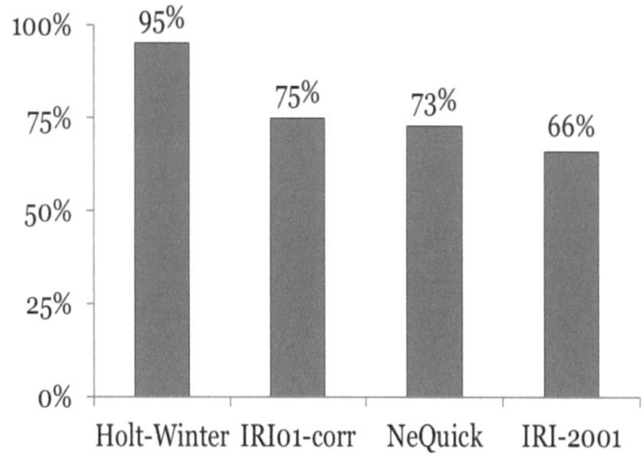

Figure 5.
The accuracy of the ionospheric models in equatorial station.

Elmunim et al. [18] compared the empirical IRI model using IRI01-corr, IRI-2001, and NeQuick topside options with the time series statical model Holt-Winter over the Langkawi, Malaysia, station in an equatorial region. Malaysia's geographic coordinates in daily, monthly, and seasonal variation, as well as during the geomagnetically disturbed period over the equatorial region are 6.19°N and 99.51°E. In comparison to the IRI01-corr, NeQuick, and Holt-Winter methods, the IRI-2001 had the greatest percentage deviation. Therefore, as shown in **Figure 5**, the model's accuracy was discovered to be about 95% for the Holt-Winter method, 75% for the IRI01-corr, 73% for the NeQuick, and 66% for the IRI-2001 model.

The Holt-Winter method performed better and estimated the TEC more accurately than the IRI01-corr and NeQuick, whereas the IRI-2001 performed poorly in the equatorial region over Malaysia. The comparative study helps to understand the model to further improve the accuracy of the GNSS. These studies are highly beneficial in correcting ionospheric delays and providing GNSS data correction, hence assisting in future accuracy improvements of the GNSS.

11. Conclusion

The GNSS satellites transmit signals propagate through the ionosphere. Ionospheric delay is one of the main sources of error that disrupts the accuracy of the GNSS signals as it changes the propagated signal's velocity. This delay can range from a few meters to tens of meters. Understanding the characteristics of the ionosphere and the estimation of the ionospheric delay are important when investigating the ionospheric delay and identifying an accurate model in a specific region. The model's accuracy depends on the complexity of the ionospheric structure and the propagating frequency of the radio waves. In equatorial regions, the time series model has given highly efficient results compared to empirical models. It is important to determine the most effective model to reduce GNSS errors and improve the accuracy of the GNSS.

Author details

Nouf Abd Elmunim
Department of Electrical Engineering, College of Engineering, Princess Nourah Bint Abdulrahman University, Riyadh, Saudi Arabia

*Address all correspondence to: naasmail@pnu.edu.sa

References

[1] Jensen AB, Mitchell C. GNSS and the ionosphere. GPS World. 2011;**22**(2):40-42

[2] Elmunim NA, Abdullah M. Ionosphere. In: Ionospheric Delay Investigation and Forecasting. Singapore: Springer; 2021. pp. 19-29

[3] Okoh D. GPS Modeling of the ionosphere using computer neural networks. In: Rustamov RB, Hashimov AM, editors. Multifunctional Operation and Application of GPS. London, UK: IntechOpen; 2018. DOI: 10.5772/intechopen.75087

[4] Parkinson BW, Enge PK, Spilker JJ. Differential gps. Progress in Astronautics and Aeronautics: Global Positioning System Theory and Applications. 1996;**2** (164):3-32

[5] Khan SZ, Mohsin M, Iqbal W. On GPS spoofing of aerial platforms: A review of threats, challenges, methodologies, and future research directions. PeerJ Computer Science. 2021;**7**:e507. DOI: 10.7717/peerj-cs.507

[6] Karaim MO, Karamat TB, Noureldin A, Tamazin M, Atia MM. Real-time cycle-slip detection and correction for land vehicle navigation using inertial aiding. In: Proceedings of the 26th International Technical Meeting of the Satellite Division of the Institute of Navigation (ION GNSS+ 2013). 2013. pp. 1290-1298

[7] Van Dierendonck AJ, Hegarty C, Scales W, Ericson S. Signal specification for the future GPS civil signal at L5. In: Proceedings of the IAIN World Congress and the 56th Annual Meeting of the Institute of Navigation (2000). 2000, June. pp. 232-241

[8] Betz JW. Effect of linear time-invariant distortions on RNSS code tracking accuracy. In: Proceedings of the 15th International Technical Meeting of the Satellite Division of the Institute of Navigation (ION GPS 2002). 2002. pp. 1636-1647

[9] Zhang J, Lachapelle G. Precise estimation of residual tropospheric delays using a regional GPS network for real-time kinematic applications. Journal of Geodesy. 2001;**75**(5):255-266

[10] Chen H, Li W, Liu X, Jiao W. A study on measuring channel bias in GNSS receiver. In: China Satellite Navigation Conference (CSNC) 2015 Proceedings. Berlin, Heidelberg: Springer; 2015. pp. 343-352

[11] Håkansson M, Jensen AB, Horemuz M, Hedling G. Review of code and phase biases in multi-GNSS positioning. GPS Solutions. 2017;**21**(3):849-860

[12] Mirmohammadian F, Asgari J, Verhagen S, Amiri-Simkooei A. Multi-GNSS-weighted interpolated tropospheric delay to improve long-baseline RTK positioning. Sensors. 2022; **22**(15):5570

[13] El-Rabbany A. GPS Positioning Modes, Introduction to GPS: The Global Positioning System. Norwood. MA: Artech House. Inc; 2002. pp. 69-83

[14] Guo Z, Yu X, Hu C, Jiang C, Zhu M. Research on linear combination models of BDS multi-frequency observations and their characteristics. Sustainability. 2022;**14**(14):8644

[15] Nade DP, Potdar SS, Pawar RP. Study of equatorial plasma bubbles using ASI and GPS systems. In: Geographic Information Systems in Geospatial Intelligence. London, UK: IntechOpen; 2020

[16] Elmunim NA, Abdullah M, Bahari SA. Evaluating the performance of IRI-2016 using GPS-TEC measurements over the equatorial region. Atmosphere. 2021; **12**(10):1243

[17] Elmunim NA, Abdullah M. Ionospheric delay forecasting. In: Ionospheric Delay Investigation and Forecasting. Singapore: Springer; 2021. pp. 41-71

[18] Elmunim NA, Abdullah M, Hasbi AM, Bahari SA. Comparison of GPS TEC variations with Holt-winter method and IRI-2012 over Langkawi, Malaysia. Advances in Space Research. 2017;**60**(2): 276-285

[19] Abd Elmunim N, Abdullah M, Bahari SA. Characterization of ionospheric delay and forecasting using GPS-tec over equatorial region. Annals of Geophysics. 2020;**63**(2):PA211

[20] Antia HM, Basu S. Local helioseismology using ring diagram analysis. Astronomische Nachrichten: Astronomical Notes. 2007;**328**(3–4): 257-263

[21] Bagiya MS, Joshi HP, Iyer KN, Aggarwal M, Ravindran S, Pathan BM. TEC variations during low solar activity period (2005–2007) near the equatorial ionospheric anomaly crest region in India. In: Annales Geophysicae. Copernicus GmbH; 2009;**27**:1047-1057

[22] Tariku YA. TEC prediction performance of the IRI-2012 model over Ethiopia during the rising phase of solar cycle 24 (2009–2011). Earth, Planets and Space. 2015;**67**(1):1-10

[23] Shimeis A, Borries C, Amory-Mazaudier C, Fleury R, Mahrous AM, Hassan AF, et al. TEC variations along an East Euro-African chain during 5th April 2010 geomagnetic storm. Advances in Space Research. 2015;**55**(9):2239-2247

[24] Akala AO, Somoye EO, Adewale AO, Ojutalayo EW, Karia SP, Idolor RO, et al. Comparison of GPS-TEC observations over Addis Ababa with IRI-2012 model predictions during 2010–2013. Advances in Space Research. 2015;**56**(8):1686-1698

[25] Le Huy M, Amory-Mazaudier C, Fleury R, Bourdillon A, Lassudrie-Duchesne P, Thi LT, et al. Time variations of the total electron content in the southeast Asian equatorial ionization anomaly for the period 2006–2011. Advances in Space Research. 2014;**54**(3): 355-368

[26] Galav P, Dashora N, Sharma S, Pandey R. Characterization of low latitude GPS-TEC during very low solar activity phase. Journal of Atmospheric and Solar-Terrestrial Physics. 2010;**72** (17):1309-1317

[27] Kumar S, Tan EL, Razul SG, See CMS, Siingh D. Validation of the IRI-2012 model with GPS-based ground observation over a low-latitude Singapore station. Earth, Planets and Space. 2014;**66**(1):1-10

[28] Sharma SK, Singh AK, Panda SK, Ansari K. GPS derived ionospheric TEC variability with different solar indices over Saudi Arab region. Acta Astronautica. 2020;**174**:320-333

[29] Li M, Yuan Y, Zhang X, Zha J. A multi-frequency and multi-GNSS method for the retrieval of the ionospheric TEC and intraday variability of receiver DCBs. Journal of Geodesy. 2020;**94**(10):1-14

[30] Sivavaraprasad G, Ratnam DV. Performance evaluation of ionospheric time delay forecasting models using GPS observations at a low-latitude station.

Advances in Space Research. 2017;**60**(2): 475-490

[31] Tariq MA, Shah M, Ulukavak M, Iqbal T. Comparison of TEC from GPS and IRI-2016 model over different regions of Pakistan during 2015–2017. Advances in Space Research. 2019;**64**(3): 707-718

[32] Feltens J, Angling M, Jackson-Booth N, Jakowski N, Hoque M, Hernández-Pajares M, et al. Comparative testing of four ionospheric models driven with GPS measurements. Radio Science. 2011; **46**(06):1-11

[33] Wu X, Zhou J, Tang B, Cao Y, Fan J. Evaluation of COMPASS ionospheric grid. GPS Solutions. 2014;**18**(4):639-649

[34] Klobuchar JA. Ionospheric time-delay algorithm for single-frequency GPS users. IEEE Transactions on Aerospace and Electronic Systems. 1987; **3**:325-331

[35] Zhang Q, Liu Z, Hu Z, Zhou J, Zhao Q. A modified BDS Klobuchar model considering hourly estimated night-time delays. GPS Solutions. 2022;**26**(2):1-13

[36] Angrisano A, Gaglione S, Gioia C, Massaro M, Robustelli U. Assessment of NeQuick ionospheric model for Galileo single-frequency users. Acta Geophysica. 2013;**61**(6):1457-1476

[37] Hoque MM, Jakowski N, Orús-Pérez R. Fast ionospheric correction using Galileo Az coefficients and the NTCM model. GPS Solutions. 2019;**23**(2):1-12

[38] Zhang X, Ma F, Ren X, Xie W, Zhu F, Li X. Evaluation of NTCM-BC and a proposed modification for single-frequency positioning. GPS Solutions. 2017;**21**(4):1535-1548

[39] Li Z, Wang N, Wang L, Liu A, Yuan H, Zhang K. Regional ionospheric TEC modeling based on a two-layer spherical harmonic approximation for real-time single-frequency PPP. Journal of Geodesy. 2019;**93**(9):1659-1671

[40] Ren X, Chen J, Li X, Zhang X, Freeshah M. Performance evaluation of real-time global ionospheric maps provided by different IGS analysis centers. GPS Solutions. 2019;**23**(4):1-17

[41] Chen J, Ren X, Zhang X, Zhang J, Huang L. Assessment and validation of three ionospheric models (IRI-2016, NeQuick2, and IGS-GIM) from 2002 to 2018. Space. Weather. 2020;**18**(6): e2019SW002422

[42] Elmunim NA, Abdullah M. Ionospheric Delay Investigation and Forecasting. Singapore: Springer; 2021

[43] Böhm J, Schuh H, editors. Atmospheric Effects in Space Geodesy (Vol. 5). Berlin: Springer; 2013

[44] Bilitza D, Altadill D, Zhang Y, Mertens C, Truhlik V, Richards P, et al. The international reference ionosphere 2012–a model of international collaboration. Journal of Space Weather and Space Climate. 2014;**4**:A07

[45] Bilitza D. IRI the international standard for the ionosphere. Advances in Radio Science. 2018;**16**:1-11

[46] Bilitza D. International reference ionosphere 2000. Radio Science. 2001;**36** (2):261-275

[47] Mallika L, Ratnam DV, Raman S, Sivavaraprasad G. Machine learning algorithm to forecast ionospheric time delays using global navigation satellite system observations. Acta Astronautica. 2020;**173**:221-231

[48] Pham TTH, Vu XH, Dien ND, Trang TT, Van Truong N, Thanh TD, et al. The

structural transition of bimetallic Ag–Au from core/shell to alloy and SERS application. RSC Advances. 2020;**10**(41): 24577-24594

[49] Yang H, Kang SJ. Exploring the Korean adolescent empathy using the interpersonal reactivity index (IRI). Asia Pacific Education Review. 2020;**21**(2): 339-349

[50] Li W, Li F, Shum CK, Shu C, Ming F, Zhang S, et al. Assessment of contemporary Antarctic GIA models using high-precision GPS time series. Remote Sensing. 2022;**14**(5):1070

[51] Erdoğan H, Arslan N. Identification of vertical total electron content by time series analysis. Digital Signal Processing. 2009;**19**(4):740-749

[52] Li D, Han M, Wang J. Chaotic time series prediction based on a novel robust echo state network. IEEE Transactions on Neural Networks and Learning Systems. 2012;**23**(5):787-799

[53] Elmunim NA, Abdullah M, Hasbi AM. Improving ionospheric forecasting using statistical method for accurate GPS positioning over Malaysia. In: 2016 International Conference on Advances in Electrical, Electronic and Systems Engineering (ICAEES). Malaysia: IEEE; 2016. pp. 352-355

[54] Iyer S, Mahajan A. Short-term adaptive forecast model for TEC over equatorial low latitude region. Dynamics of Atmospheres and Oceans. 2022;**2022**: 101347

[55] Suwantragul S, Rakariyatham P, Komolmis T, Sang-In A. A modelling of ionospheric delay over Chiang Mai province. In: Proceedings of the 2003 International Symposium on Circuits and Systems. ISCAS'03. Thailand: IEEE; 2003. pp. II-II

[56] Gelper S, Fried R, Croux C. Robust forecasting with exponential and Holt–Winters smoothing. Journal of Forecasting. 2010;**29**(3):285-300

[57] Elmunim, N. A., Abdullah, M., Hasbi, A. M., & Bahari, S. A. (2015). Comparison of statistical Holt-Winter models for forecasting the ionospheric delay using GPS observations. 94.20. Cf; 91.10. Fc; 95.75. Wx.

Section 2

Modeling Ionosphere

Chapter 5

Seasonal Characteristics of the Rate of Ionospheric TEC Index in China Based on BDSGAS from 2019 to 2022

Chengli She

Abstract

The ionosphere total electron content (TEC) has always been one of the important error sources in satellite-based navigation and positioning applications. Besides TEC, the rate of TEC index (ROTI) which is derived from TEC, is also served as a significant parameter in monitoring the status of the ionosphere. Based on more than 2,800 GNSS stations in China, as part of the Beidou Satellite Ground Augmentation System (BDSGAS), the ROTI maps were constructed from 2019 to 2022, with the temporal resolution of 300 seconds and the spatial resolution of 0.5°and 0.25° along longitude and latitude, respectively. In order to analyze the seasonal characteristics of ROTI maps during those years, the geomagnetic quiet days in each season are chosen based on hourly geomagnetic Dst index. Among those ROTI maps, the seasonal characteristics are obvious and differ along latitude, in the low latitude the strongest ROTI occurring and related to the scintillations in both spring and autumn, in the middle latitude ROTI active only in summer, and in the high latitude ROTI displaying no activity during each winter.

Keywords: Total Electron Content (TEC), Rate of TEC Index (ROTI), scintillation, Beidou Satellite Ground Augmentation System (BDSGAS), ionosphere

1. Introduction

In recent years, the pattern of global satellite navigation systems is undergoing great changes. China's self-developed Beidou Satellite Navigation and Positioning System (BDS), officially opened on July 3, 2020, becomes the third mature satellite navigation system after GPS and GLONASS. The comprehensive completion of BDS indicates that China has an independent voice and provides an opportunity in the field of satellite navigation. At the same time, relevant researches and applications could also be accelerated and upgraded.

At present, people's needs for navigation and positioning are becoming more and more diversified, generating various application scenarios such as automatic driving,

UAV operation, safety monitoring, etc. Among those applications, GNSS satellite signal is an indispensable signal source. However, the ionosphere, as the medium for the radio-wave transmitting from GNSS satellite to ground equipment, can never be ignored due to its refraction, scattering, and reflection upon radio signals. Therefore, to investigate the ionospheric influences based on the GNSS signals becomes an important problem to be solved. Traditionally, the ionospheric total electron content (TEC) can be inversed from double-frequency GNSS signals by integrating the ionospheric electron density along the propagation path, which is not only an important parameter in ionospheric physics research, but also an important error source to be eliminated in navigation and positioning applications.

Besides TEC, the TEC Rate of Change Index (ROTI) derived from TEC, is also frequently adopted to monitor the ionospheric state. Meanwhile, TEC and ROTI maps are favored by researchers and engineers in order to display the difference of overall regions. TEC maps are much more popular and widely-used than ROTI maps. So far, there are several research institutions regularly releasing their global ionospheric maps (GIM), such as Center for Orbit Determination in Europe (CODE, Switzerland), Jet Propulsion Laboratory (JPL, USA), European Space Agency (ESA, Germany), and Universitat Politècnica de Catalunya (UPC, Spain), Chinese Academy of Sciences (CAS, China) [1–4]. Those GIMs are submitted to International GNSS Service (IGS) to generate the combined version [5].

Particularly, in the Arctic and Antarctic, low latitudes and equatorial regions, where the ionosphere tends to change intensely, ROTI map is often used to locate the position of the active area of the ionosphere. As early as 1997, Pi et al. [6] obtained the global ROTI map in the form of scattered points with the help of global GPS data, and used it to analyze the global distribution of ionospheric irregularities. In Europe, due to the influence of aurora, Cherniak et al. [7] used more than 700 GNSS stations in the middle and high latitudes of the Northern Hemisphere to develop the Northern Hemisphere high latitude ROTI map in the geomagnetic coordinate system (the geomagnetic latitude is above 50°), whose resolutions along the magnetic latitude and magnetic local time are 2° and 8 minutes, respectively, which can easily indicate the regions with active TEC changes under the influence of the aurora. In India, Harsha et al. [8] utilized 26 Indian GNSS stations to build TEC and ROTI maps to monitor the status of the ionosphere in the equator and low latitude regions.

In China, especially before the Beidou system was fully built, TEC estimations generally rely on GPS and GLONASS signals, and the number of GNSS stations used for ionospheric monitoring is not as dense as it is now. In low latitude of southern China, the lack of observation often leads to inaccurate estimation of the ionospheric TEC, then resulting inaccurate TEC maps. Meanwhile, the researches about ROTI are mainly focus on two aspects, one is the conventional observations of ROTI in the low latitude area of southern China, and the other is the case studies on the ionospheric ROTI response under the condition of geomagnetic activities. The ROTI maps have not been regularly released and studied, whose climatological characteristics are still not clear on the overall scope of China. It is urgent to fill the gap.

In this chapter, based on the GNSS observations from over 2800 Beidou Satellite Ground Augmentation System (BDSGAS) stations in China, a set of ionospheric ROTI maps will be constructed from 2019 to 2022, and only the ones under geomagnetic quiet condition will be adopted to statistically analyze its characteristics, including the features along each season and the low, middle, and high latitudes.

2. Data and method

The GNSS observations used in this chapter are from the reference stations of BDSGAS [9]. The system adopts advanced system architecture, data processing system, software and multiple broadcasting means to enhance the positioning accuracy by multi-GNSS and multi-modes. The enhanced data products can be broadcasted by multiple means with various precisions such as meter, decimeter, and centimeter level for real-time service and even millimeter level for the post-processing positioning service. The acceptance of the system was completed in 2019. Currently, this system is operated by Qianxun Spatial Intelligence Co., Ltd., established jointly by China North Industries Group Corporation Limited and Alibaba Group in August 2015. As successful business cases, a group of customized products and services for specific application scenarios has been launched, such as dangerous building detection, precision agriculture, and autonomous vehicles.

Up to now, the system has established more than 3000 GNSS reference stations in China, covering all provinces, cities, and major counties. Each station is equipped with multi-GNSS receiver which fully supports BDS, GPS, Galileo, and GLONASS signals, with the sampling rate as high as 1Hz. Those stations can provide sufficient GNSS data for the construction of the ROTI map in this chapter.

In order to measure the level of the geomagnetic disturbance, the geomagnetic Dst index is chosen from the World Data Center (WDC) in Tokyo, Japan. According to the trend of the Dst index, the geomagnetic quiet days are selected from every month, acting as the reference for the geomagnetic disturbances. When analyzing the seasonal climatological characteristics of ROTI, only the geomagnetic quiet days during the equinox/solstice months are considered for simplicity, i.e. March, June, September, and December, respectively. The time span covers from spring 2019 to winter 2022, nearly four years long. **Table 1** lists each selected geomagnetic quiet day and the daily minimum of hourly Dst index. The minimum hourly Dst of those quiet days is no less than −10 nT.

The process of ROTI map construction can be divided into three steps. At first, ionospheric slant TEC along the ray paths from each GNSS station to satellites is inversed. The algorithm about TEC inversion is very mature, and the method based on local spherical symmetry and thin layer model assumption in literature [10] is chosen. The ionospheric puncture point (IPP) is 450 km above the Earth surface. Secondly, Rate of TEC (ROT) and ROTI for each station is obtained with the common method [7]. It should be noted that, the cut-off limit of ground elevation is set as 30°, in order to avoid the impact of multipath effects caused by low elevation and obtain the most reliable ROT and ROTI information. At last, all the ROTI data are gridded into a two-dimensional map divided by longitude and latitude [11], with the longitude and latitude resolution of 0.5° and 0.25°, respectively, and the temporal resolution is 300 seconds. The value of ROTI in a grid is the averaged ROTI of all the valid ROTI data from the ground-satellite ray paths crossing through the grid, without any additional interpolation applied.

After the above three steps, the ROTI map is completed by the form of grid. Due to the uneven distribution of GNSS stations, there are some empty grids in the map. In fact, the ROTI map is only a partially filled grid map.

3. Seasonal characteristics

On the one hand, according to the geographic latitude, the whole China and nearby region can be divided into three parts: (a) the low latitude region, located below 25°N, mainly including Hainan, Guangdong, Guangxi, and Yunnan; (b) the middle latitude district, between 25°N and 45°N, such as Sichuan, Hubei, Henan, Shandong, and Hebei; (c) the high latitude area, above 45°N, mainly including Heilongjiang, Jilin, northern Xinjiang, and northern Inner Mongolia. On the other hand, due to the fact that ROTI is more active during night than that in white day, only the ROTI maps at nighttime (18:00–03:00 LT) are considered. Then, the nighttime ROTI maps are compared by each season and each region, in order to discuss the seasonal characteristics of ROTI maps from 2019 to 2022.

3.1 Spring

Ionospheric scintillation in low latitude areas is common phenomenon in spring. For example, Li Guozhu confirmed that ionospheric scintillation is more likely to occur in spring and autumn but less likely to happen in winter and summer by using the ionospheric GPS amplitude scintillation data of Sanya station in Hainan from 2004 to 2006 [12]. Liu K. utilized the GPS data from Sanya in China from 2004 to 2012 and found that the scintillations have a maximum occurrence during equinox of solar maximum [13]. Considering the close relationship between scintillation and ROTI, the value of ROTI in spring tends to be higher in low latitude than that in middle and high latitudes.

As a group of typical examples listed in **Table 1**, the geomagnetic quiet days in spring from 2019 to 2022 are March 22, 2019, March 11, 2020, March 11, 2021, and March 22, 2022, respectively, with the hourly Dst index no less than −1 nT. The corresponding nighttime ROTI maps for each day are shown in **Figure 1**. It can be seen that ROTI is less than 0.1 TECU/min in most of the area from 2019 to 2022, expect in

season	Date of the geomagnetic day	Minimum of hourly Dst (nT)
Spring	2019-03-22	−1
	2020-03-11	4
	2021-03-11	6
	2022-03-22	−1
Summer	2019-06-12	−1
	2020-06-15	−6
	2021-06-14	−7
	2022-06-13	−4
Autumn	2019-09-19	−8
	2020-09-10	−3
	2021-09-12	−7
	2022-09-18	−7
Winter	2019-12-17	−8
	2020-12-14	−6
	2021-12-15	−10
	2022-12-18	−1

Table 1.
Geomagnetic quiet days and daily minimum of hourly Dst indices during 2019 to 2022.

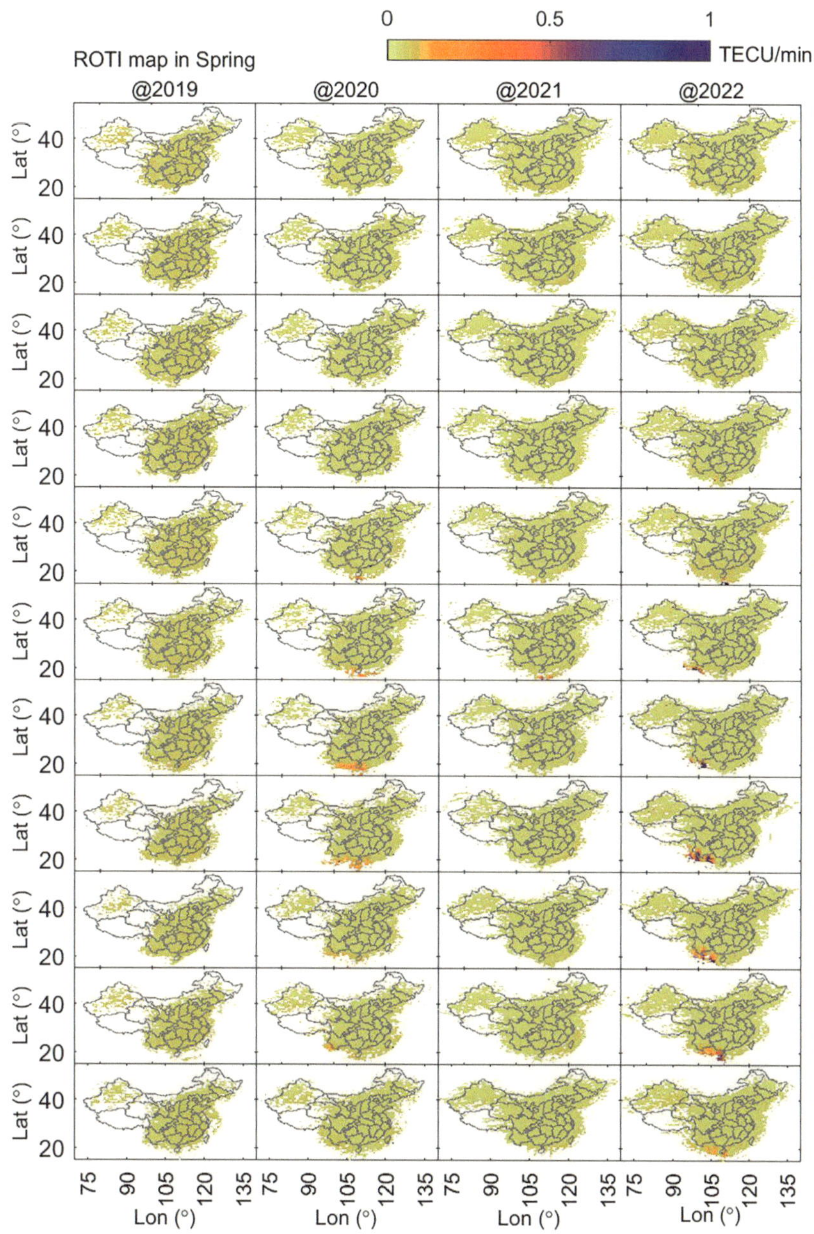

Figure 1.
The nighttime ROTI maps in spring.

the low latitude of Hainan and nearby region, where the geographic latitude is less than 20°N. Since the value of ROTI acts as the sign of scintillations to some extent, the higher the ROTI is, the stronger the scintillation is. The case of ROTI less than 0.1 TECU/min can be marked as "blank" without any scintillation. Therefore, there is nearly no scintillation in 2019 spring, the quietest spring among those years. In 2020 spring, scintillations were obvious only in Hainan and around during 22:00 LT to 02:00 LT, below 20°N. In 2021 spring, scintillations are observed only far south of Hainan around 23:00 LT, also below 20°N but with much smaller size than those in

2020 spring. In 2022 spring, the scintillations occurred below 20°N from 22:00 LT to 04:00 LT, lasting longer than other years.

The strength of scintillation in each year is also different. To measure the strength, a group of ROTI thresholds are considered. The strength can be compared by counting the total number and calculating the average of ROTI among the grids exceeding the thresholds. For simplicity, the thresholds chosen in this chapter are 0.1 TECU/min and 0.5 TECU/min, referring to no scintillation and moderate scintillation, respectively. Three LTs are selected to evaluate the temporal difference, i.e.: 19:00 LT, 21:00 LT, and 23:00 LT. The results are shown in **Table 2**. It can be clearly seen that, 2022 spring is the strongest among those years, with the most number and the largest average of grid exceeds 0.5 TECU/min, while 2019 spring is the weakest. In 2020 and 2021 spring, scintillations are almost equivalent in strength. The difference of such strength from 2019 to 2022 may be related to the different level of solar activity, increasing from solar minimum to growing phase.

	LT	Thresholds (TECU/min)	Number	Average (TECU/min)
2019	19:00	0.1	156	0.193
		0.5	4	2.902
	21:00	0.1	108	0.328
		0.5	6	3.764
	23:00	0.1	212	0.287
		0.5	13	2.850
2020	19:00	0.1	94	0.142
		0.5	2	1.177
	21:00	0.1	137	0.140
		0.5	2	1.117
	23:00	0.1	364	0.230
		0.5	22	0.909
2021	19:00	0.1	66	0.143
		0.5	2	0.712
	21:00	0.1	102	0.175
		0.5	4	1.023
	23:00	0.1	177	0.330
		0.5	38	0.838
2022	19:00	0.1	91	0.323
		0.5	10	1.823
	21:00	0.1	81	0.342
		0.5	8	2.064
	23:00	0.1	225	0.757
		0.5	70	2.01

Table 2.
Number and ROTI average of grids for ROTI exceeding the thresholds in spring.

3.2 Summer

Different from the ionospheric scintillation mainly in the low latitude region in spring, the ionospheric nighttime activities in summer often correspond to the ionospheric electron density depletions caused by the mid latitude ionospheric trough (MIT) phenomenon [14]. For example, in the middle latitude of East Asia, He Youwen and Long Qili analyzed the seasonal characteristics of the ionospheric scintillation

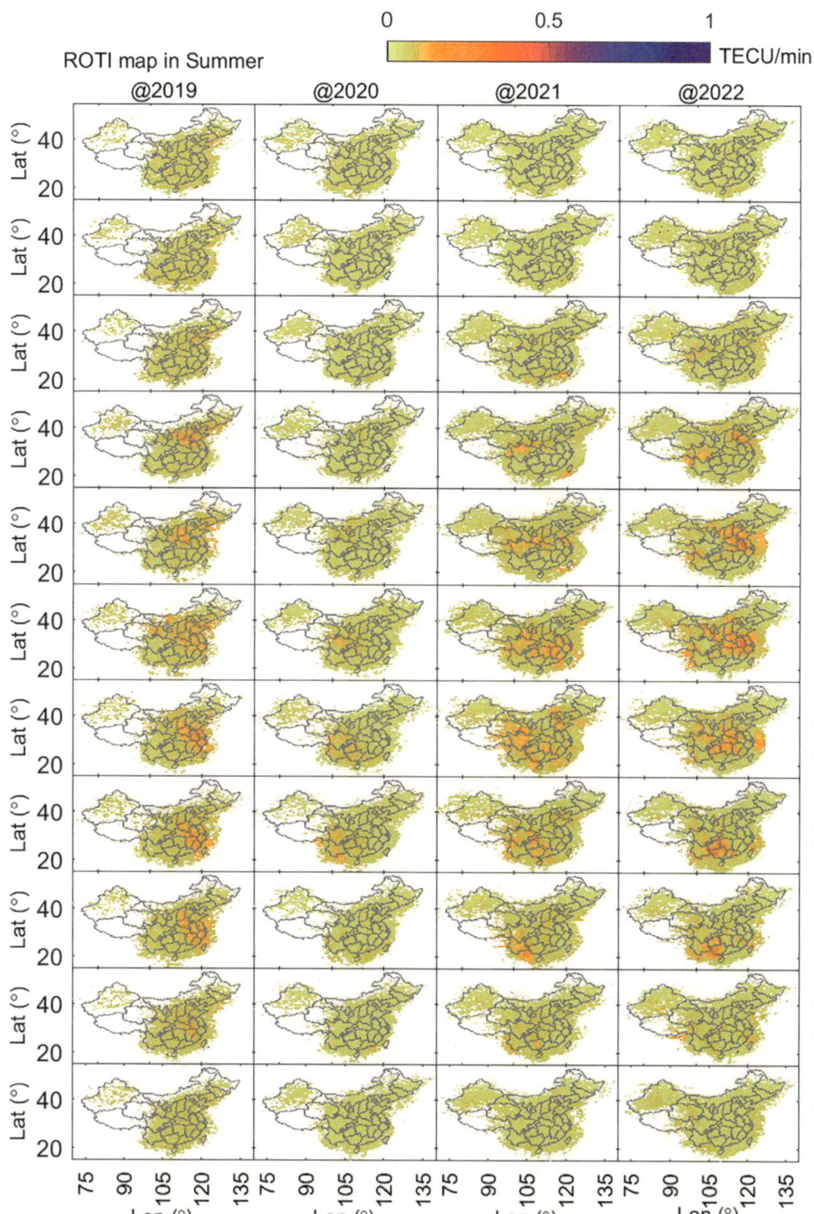

Figure 2.
The nighttime ROTI maps in summer.

from the scintillation observations of four stations, and found that the occurrence rate of scintillation is highest in summer, followed by autumn and weakest in winter [15]. However, due to the restriction of observation conditions, there are few reports using ROTI maps to study this phenomenon.

June 12, 2019, June 15, 2020, June 14, 2021, and June 13, 2022 are chosen as the geomagnetic quiet days in the summer of each year, as listed in **Table 1** all the hourly Dst indices in summer greater than −9 nT. The nighttime ROTI maps are shown in **Figure 2**. It can be seen from the figure that, in the night of each summer, ROTI greater than 0.1 TECU/min is widespread in mid latitude regions during 21:00 LT to 02:00 LT, but rarely occurs in low and high latitudes, with the size of active ROTI much wider than that in spring. Meanwhile, as time goes on, the active region of ROTI larger than 0.1TECU/min in the map moves from east to west, which is mainly related to the Earth's rotation resulting in the movement of direct solar radiation point from east to west.

	LT	Thresholds (TECU/min)	Number	Average (TECU/min)
2019	19:00	0.1	479	0.189
		0.5	10	3.269
	21:00	0.1	1002	0.138
		0.5	3	1.919
	23:00	0.1	1575	0.142
		0.5	9	2.768
2020	19:00	0.1	70	0.129
		0.5	1	0.555
	21:00	0.1	109	0.165
		0.5	2	2.398
	23:00	0.1	377	0.123
		0.5	0	—
2021	19:00	0.1	109	0.216
		0.5	3	3.267
	21:00	0.1	581	0.127
		0.5	2	0.952
	23:00	0.1	1603	0.124
		0.5	0	—
2022	19:00	0.1	144	0.183
		0.5	6	1.214
	21:00	0.1	743	0.134
		0.5	4	0.989
	23:00	0.1	2023	0.153
		0.5	10	1.472

Table 3.
Number and ROTI average of grids for ROTI exceeding the thresholds in summer.

As the same ROTI thresholds used in spring, the total number and the average of ROTI among the grids exceeding the thresholds are listed in **Table 3**. Since the number of grids exceeding 0.5 TECU/min is too small, such grids can be considered as gross error to be ignored. The situation in summer is quite different from that in spring. In summer nighttime, the number of grids exceeding 0.1 TECU/min can reach 2023 around 23:00 LT in year 2022, but in spring only 364 grids are captured, much more than that in spring, spreading much more widely in summer. The averages of ROTI for grids over 0.1 TECU/min in summer are generally from 0.123 TECU/min to 0.216 TECU/min, but the same averages in spring are from 0.142 TECU/min to 0.757 TECU/min, less in summer than that in spring. From 2019 to 2022, the number of grids exceeding 0.1 TECU/min is smallest in 2020, and largest in 2022, also following different pattern of activity other than that in spring. Even though the solar activity reached its minimum during 2019–2020, much less active than in 2021, the nighttime ROTI map in summer is still as strong as in 2021, nearly uninfluenced by the level of solar activities.

3.3 Autumn

Similar to spring, autumn is also another season with high occurrence of ionospheric scintillations in low latitudes. Zhang Hongbo et al. [10] used the ionospheric scintillation data of Haikou Station in China's low latitude region from 2010 to 2017 to analyze the statistical occurrence of ionospheric scintillation in spring and autumn, found that both the occurrence and the intensity of ionospheric scintillation were generally higher in spring equinox than those in autumn equinox, and attributed the inconsistency between spring and autumn to the asymmetry of the equinox in the ionospheric background.

As listed in **Table 1**, the geomagnetic quiet days in each autumn from 2019 to 2022 are September 19, 2019, September 10, 2020, September 12, 2021, and September 18, 2022, respectively, with the hourly Dst index no less than −8 nT. The nighttime ROTI maps are shown in **Figure 3**. Seen from **Figure 3**, the scintillations were almost blank both in 2019 and 2020 autumn, occurred only in a tiny small area of the southeast region in 2021 autumn from 22:00 LT to 00:00 LT, and extended to larger area in 2022 autumn than those in 2021 from 21:00 LT to 03:00 LT. The intensity of scintillations in each autumn can also be clearly revealed, the strongest in 2022, followed by 2021, and the weakest in 2020.

The total number and the average of grids exceeding the ROTI thresholds of 0.1 TECU/min and 0.5 TECU/min are listed in **Table 4**. As the strongest autumn in 2022 among those years, around 23:00 LT in 2022, the number and the ROTI average of grids over 0.5 TECU/min is 140 and 1.398 TECU/min, respectively, while the number of grids over 0.5 TECU/min in other years is no more than 11. The ROTI average of grids over 0.1 TECU/min is a little larger in 2021 than those in both 2019 and 2020. When comparing spring and autumn of the same year, the number of grids over the thresholds is usually larger in spring than that in autumn except in 2022 and 19:00–21:00LT in 2019, while the ROTI average is also larger in spring than that in autumn except the gross error with only several scattering grids over 0.5 TECU/min. Generally, both the range and amplitude of scintillations are weaker in autumn than those in spring. Therefore, the difference of scintillations in spring and autumn in each year is consistent with the performance of ionospheric scintillation stronger in spring than autumn as mentioned in the literature [16].

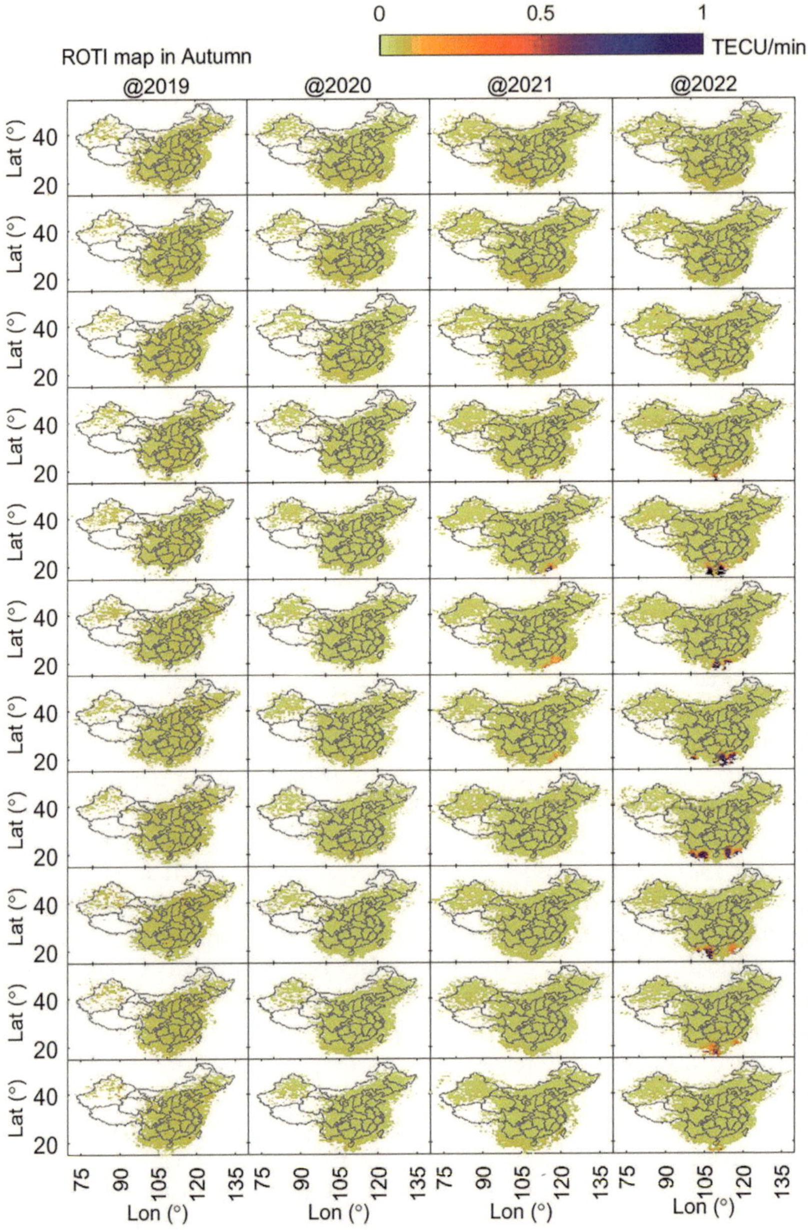

Figure 3.
The nighttime ROTI maps in autumn.

3.4 Winter

In winter, the direct sunlight point moves to the southern hemisphere, thus the ionosphere changes in the northern hemisphere are relatively stable and without scintillation at night. The geomagnetic quiet days for each winter from 2019 to 2022 are December 17, 2019, December 14, 2020, December 15, 2021, and December 18, 2022, respectively, with the minimal value of hourly Dst index no less than −10 nT.

	LT	Thresholds (TECU/min)	Number	Average (TECU/min)
2019	19:00	0.1	175	0.155
		0.5	4	1.527
	21:00	0.1	153	0.216
		0.5	11	1.396
	23:00	0.1	155	0.140
		0.5	3	1.185
2020	19:00	0.1	159	0.120
		0.5	0	—
	21:00	0.1	89	0.137
		0.5	1	1.887
	23:00	0.1	33	0.174
		0.5	1	1.757
2021	19:00	0.1	87	0.167
		0.5	1	3.992
	21:00	0.1	109	0.240
		0.5	5	2.208
	23:00	0.1	170	0.166
		0.5	2	0.689
2022	19:00	0.1	94	0.183
		0.5	6	0.765
	21:00	0.1	248	0.494
		0.5	48	1.909
	23:00	0.1	346	0.686
		0.5	140	1.398

Table 4.
Number and ROTI average of grids for ROTI exceeding the thresholds in autumn.

The corresponding nighttime ROTI maps are shown in **Figure 4**. Different from all the ROTI maps in other seasons, the nighttime ROTI maps always keep in very low level in each winter, without any active area. Therefore, there is no scintillation in winter. Then, the analysis about the ROTI thresholds is also omitted in this subsection.

4. Conclusions

Under geomagnetic quiet conditions, the nighttime ROTI maps for each season from 2019 to 2022 in China is constructed, and its seasonal characteristics are analyzed in this chapter. The following conclusions can be drawn: (1) In spring and autumn, ROTI is mainly active only in the low latitude areas in southern China. The value of ROTI in active areas is larger than that in other seasons, which is also related to ionospheric scintillations; (2) In summer, ROTI is active only in the middle latitude of central China, with the amplitude of ROTI less than that in Equinox. The size of active ROTI area is much wider than that in spring and autumn. It is generally related to the ionospheric nighttime trough in summer; (3) In winter, the ionosphere is generally quiet, without any active ROTI area.

In addition to the ROTI map mentioned in this chapter, the ionospheric scintillation S4 index is also often used for ionospheric state monitoring. In the future, more

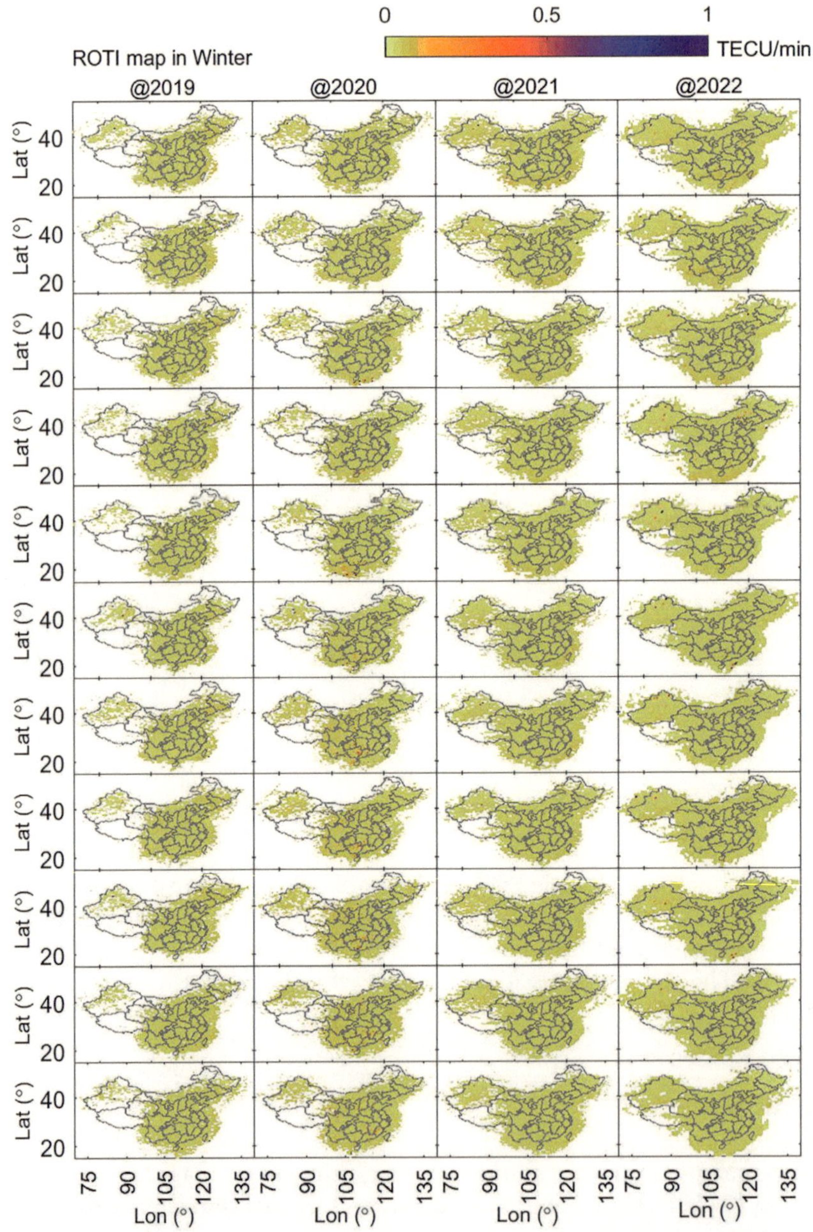

Figure 4.
The nighttime ROTI maps in winter.

comparative studies about various parameters can be carried out to provide comprehensive ionospheric monitoring services.

Acknowledgements

This work was supported by the National Natural Science Foundation of China (41704158) and the National Key Research and Development Program of China (2021YFA0717300). Thanks to the World Data Center (WDC) in Tokyo, Japan for providing geomagnetic Dst data.

Conflict of interest

The author declares no conflict of interest.

Author details

Chengli She
Qianxun Spatial Intelligence (Zhejiang) Inc., Deqing, Zhejiang, China

*Address all correspondence to: snake553@163.com

References

[1] Mannucci A, Wilson BD, Yuan DN, Ho CH, Lindqwister UJ, Runge TF. A global mapping technique for GPS-derived ionospheric total electron content measurements. Radio Science. 1998;**33**(2):565-582. DOI: 10.1029/97RS02707

[2] Schaer S. Mapping and Predicting the Earth's Ionosphere Using the Global Positioning System. Berne: Astronomical Institutes, University of Bern; 1999

[3] Hernández-Pajares M, Juan JM, Sanz J. New approaches in global ionospheric determination using ground GPS data. Journal of Atmospheric and Solar—Terrestrial Physics. 1999;**61**(16):1237-1247. DOI: 10.1016/S1364-6826(99)00054-1

[4] Li Z, Yuan Y, Wang N, Hernandez-Pajares M, Huo X. SHPTS: towards a new method for generating precise global ionospheric TEC map based on spherical harmonic and generalized trigonometric series functions. Journal of Geodesy. 2015;**89**(4):331-345

[5] Hernández-Pajares M, Juan J, Sanz J, Orus R, Garcia-Rigo A, Feltens J, et al. The IGS VTEC maps: a reliable source of ionospheric information since 1998. Journal of Geodesy. 2009;**83**(3):263-275. DOI: 10.1007/s00190-008-0266-1

[6] Pi X, Mannucci AJ, Lindqwister UJ, Ho CM. Monitoring of global ionospheric irregularities using the worldwide GPS network. Geophysical Research Letters. 1997;**24**(18):2283-2286

[7] Cherniak I, Krankowski A, Zakharenkova I. Observation of the ionospheric irregularities over the Northern Hemisphere: Methodology and service. Radio Science. 2014;**49**(8):653-662

[8] Harsha PB, Ratnam DV, Nagasri ML, Sridhar M, Raju KP. Kriging-based ionospheric TEC, ROTI and amplitude scintillation index (S4) maps for India. IET Radar, Sonar and Navigation. 2020;**14**:1827-1836. DOI: 10.1049/iet-rsn.2020.0202

[9] Tianxiong L. Principle and Application of Satellite Navigation Augmentation System (in Chinese). Beijing: National Defense Industry Press; 2021

[10] She C, Yue X, Hu L, Zhang F. Estimation of ionospheric total electron content from a Multi-GNSS Station in China. IEEE Transactions on Geoscience and Remote Sensing. 2019;**58**(2):852-860

[11] She C, Liu H, Yu J, Zhou P, Cui H. Development of high-precision ionospheric monitoring system in China: Taking ROTI Map as an Example. In: China Satellite Navigation Conference (CSNC 2021); 26-28 May 2021. Nanchang, Singapore: Springer; 2021. pp. 187-196

[12] Guozhu L. Studies on Monitoring, Analysis and Application of Mid-and Low-Latitude Ionospheric Scintillation in China (in Chinese). Wuhan: Wuhan Institute of Physics and Mathematics, Chinese Academy of Sciences; 2007

[13] Liu K, Li G, Ning B, Hu L, Li H. Statistical characteristics of low-latitude ionospheric scintillation over China. Advances in Space Research. 2015;**55**(5):1356-1365

[14] He M, Liu L, Wan W, Zhao B. A study on the nighttime midlatitude ionospheric trough. Journal of Geophysical Research: Space Physics. 2011;**116**:A05315. DOI: 10.1029/2010JA016252

[15] He Y, Qili L. Mid-latitude Ionospheric Scintillation in Eastern Asia (in Chinese). Chinese Journal of Radio Science. 1994;**4**(02):45-51

[16] Hongbo Z, Wang F, Xu L, Dongsheng S, Yumei L. A Seasonal variation model of ionospheric scintillation occurrences with the equinoctial asymmetry (in Chinese). Chinese Journal of Radio Science. 2019;**34**(02):180-185

Chapter 6

Ionospheric Scintillation Models: An Inter-Comparison Study Using GNSS Data

Adriano Camps, Carlos Molina, Guillermo González-Casado, José Miguel Juan, Joël Lemorton, Vincent Fabbro, Aymeric Mainvis, José Barbosa and Raúl Orús-Pérez

Abstract

Existing climatological ionosphere models, for example, GISM, SCIONAV, WBMOD, and STIPEE, have known limitations that prevent their wide use. In the framework of ESA study "Radio Climatology Models of the Ionosphere: Status and Way Forward" their performance was assessed using experimental observations of ionospheric scintillation collected over the past years to evaluate their ability to properly support future missions, and eventually indicate their weaknesses for future improvements. Model limitations are more important in terms of the intensity scintillation parameter (S_4). To improve them, the COSMIC model has been fit (scaling factor and offset) to the measured data, and it became the one better predicting the intensity scintillation in a statistical sense.

Keywords: ionosphere, scintillation, intensity, phase, GNSS, climatology, modeling

1. Introduction

Ionospheric scintillations are the random intensity (I) and phase (σ_ϕ) fluctuations suffered by electromagnetic waves traversing the ionosphere due to irregularities of the electron content density, mostly originated by the solar activity and plasma irregularity processes. They are characterized by the amplitude (actually intensity, i.e., power) scintillation parameter (S_4) and the phase scintillation parameter (σ_ϕ):

$$S_4 = \sqrt{\frac{\langle I^2 \rangle - \langle I \rangle^2}{\langle I \rangle^2}}, \quad (1)$$

$$\sigma_\phi = \sqrt{\langle \phi^2 \rangle - \langle \phi \rangle^2}, \quad (2)$$

where I is the intensity of the signal (i.e., power), ϕ is the detrended phase of the signal, and $\langle . \rangle$ denotes the time average.

Ionospheric scintillation impacts satellite communications, global navigation satellite systems (GNSS), and remote sensors (e.g., UHF Synthetic Aperture Radars—SAR–, radar sounders, GNSS-Reflectometry, and GNSS-Radio Occultations). Ionospheric scintillation mostly occurs in equatorial and high-latitude regions, and their behavior is significantly different. The complexity of the physical processes occurring in the Earth's ionosphere-thermosphere-mesosphere system is summarized in **Figure 1a**, while **Figure 1b** shows the main ionospheric layers and the electron density profiles during day and night.

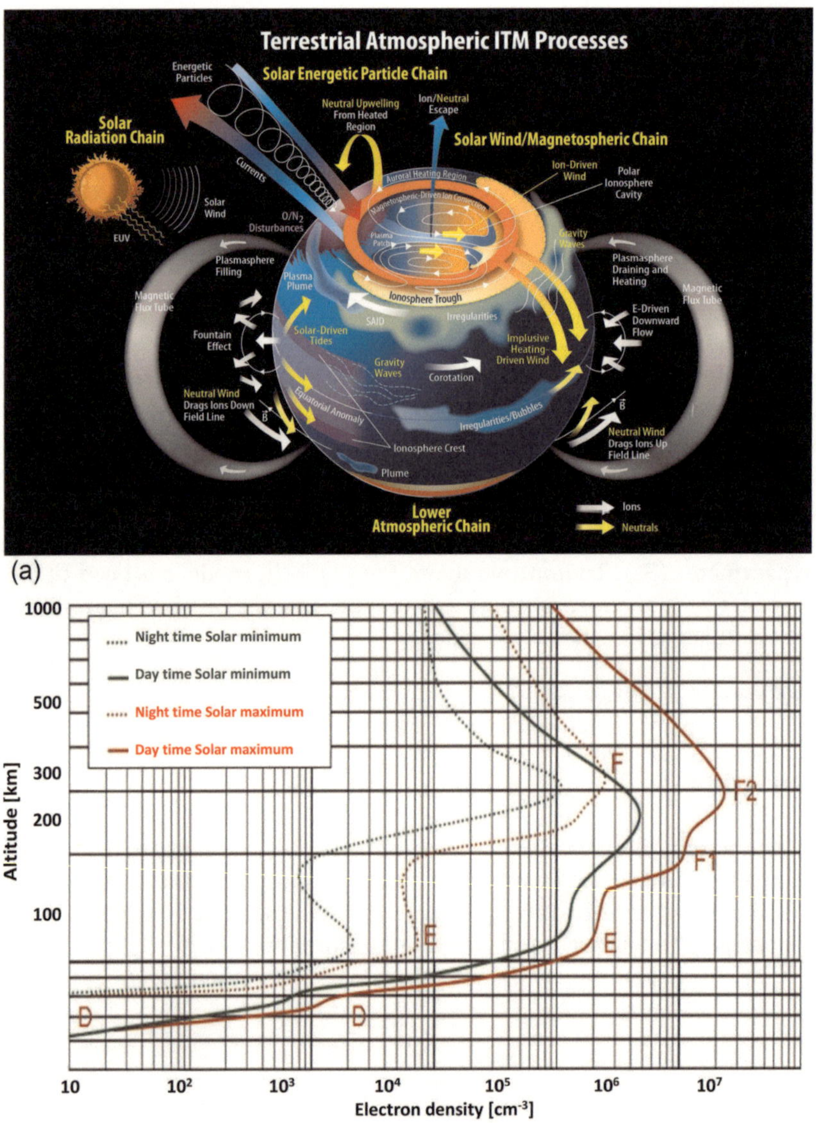

Figure 1.
(a) Indication of the complexity of the ionosphere-thermosphere-mesosphere (ITM) system of planet Earth and the range of physical processes operating. Credits: NASA's Goddard Space Flight Center/Mary Pat Hrybyk-Keith [https://svs.gsfc.nasa.gov/12960].(b) Typical electron density profile in the ionosphere and ionospheric layers during day/night (from https://sidstation.loudet.org/ionosphere-en.xhtml).

A summary of the main scintillation models is given in the excellent review presented in [1]:

Equatorial scintillations occur around ±20° of latitude of the magnetic equator, after sunset and before midnight. They are caused by small-scale structures, from tens of meters to tens of km [2], in convective plasma processes surrounding depleted ionization volumes driven through the F region, which can extend well through the F-layer peak. Irregularities with this range of scales are not independent from larger-scale plasma structures and are also related to smaller-scale irregularities.

Mid-latitude scintillations occur as an extension of phenomena occurring at equatorial and auroral latitudes. They can also be due to an intense sporadic E layer during daytime. High-latitude scintillations occur from the high-latitude edge of the Van Allen outer belt into the polar region. The greater scintillation occurrence is during the dark months, rather than during the months of continuous Sun illumination, at all local times.

Auroral zones are observed during the nighttime period. Scintillation at high latitudes is mostly refractive, and its impact in GNSS signals shows a well-known proportionality between different signal frequencies (e.g., [3]). Moreover, at high latitudes, the ionospheric disturbances mostly produce phase fluctuations, but little impact in the signal amplitude/intensity [4]. This is not the case for the low-latitude scintillation, where scintillation produces diffractive effects on the signals, and the proportionality of the effects with the signal frequency is broken and, moreover, the signal amplitude/intensity is affected (e.g., [4, 5]).

Empirical methods, such as the Basu et al. Equatorial scintillation model [6], Basu High-Latitude Scintillation Model [7], or the WAM Model [8] are restricted to geographical areas, periods of time, frequency bands in which they were derived. Analytical methods such as the Fremouw and Rino Model, the first analytical model of VHF/UHF scintillations [9], the Aarons Model [10], the Franke and Liu Model [11], the Iyer et al. Model [12], or the Retterer Model [13] assume that the ionosphere is a layer of free electrons at a given height and with a given thickness that, under the presence of the magnetic field of the Earth, disturbs the propagation of the electromagnetic waves. More recently, Liu et al. [14] and Chen et al. [15] derived an empirical model to estimate S_4 from FormoSat-3/COSMIC observations, and Kepkar et al. [16] use FormoSat-3/COSMIC data to characterize equatorial plasma bubbles.

Climatological models, include Global Climatological Models such as the WBMOD (WideBand MODel), the STIPEE, the GISM (Global Ionospheric Scintillation Model), and the SCIONAV model.[1]

- WBMOD [17–19] is a climatological model for ionospheric turbulence that includes the global distribution of the electron density irregularities and the phase screen propagation model by Rino [20, 21] to calculate the effects that density irregularities produce in the propagation of electromagnetic waves. WBMOD

[1] WBMOD software is not open, although ONERA owns a license. STIPEE is a proprietary software of ONERA. GISM is open and it is the one adopted by the International Telecommunications Union. An online tool of GISM exists at [http://www.ieea.fr/en/gism-web-interface.html]. SCIONAV was developed for ESA and it is open under request.

provides modeling of the integrated strength of inhomogeneity (C_kL), the medium velocity, the anisotropy, the slope (p), and the outer scale of turbulence. Its outputs are the phase scintillation spectrum spectral index, the spectral strength parameter (at 1 Hz) T, the intensity scintillation index S_4 (Eq. (1)), and the rms of detrended phases σ_ϕ [in radians] (Eq. (2)). Assuming that the power spectral density of the phase fluctuations is isotropic, it can be approximated by:

$$\gamma_\phi(k) = \frac{C_s}{\left(k^2 + q_0^2\right)^{p/2}}, \tag{3}$$

where C_S characterizes the turbulence strength, k is the wave number, $q_0 = 2\pi/L_0$ being L_0 the outer scale of the inhomogeneities, and p is the spectrum slope (the spectrum is linear in log–log scale).

WBMOD gives the probability distribution of the scintillation levels for any position at any time: at high latitudes, a single Gaussian probability density function (PDF) describes the $\log(C_kL)$, at low latitudes, a bimodal representation of the PDF is proposed, corresponding to the occurrence of plumes, and specific models for polar regions are included for the auroral oval and medium velocity.

Its main limitations are that the predicted scintillation activity is much lower than that observed and this disagreement increases for stronger scintillation, that the scintillation activity predicted by WBMOD ceases approximately 2 h earlier than the observations show, and that the patchy character of the equatorial scintillations is not reflected in the model [22]. Also, the spectrum slope (p) is fixed to 2.5 or 2.7, which prevents the parametrization of the ionospheric turbulence spectrum in polar regions (see **Figure 2a** and **b** in Section 3.3). Additionally, WBMOD was obtained from a large set of low-frequency beacon measurements (Wideband, HiLat, and Polar BEAR satellite experiments, USAF Phillips Laboratory equatorial scintillation monitoring network), therefore its validity at the GNSS frequency bands that will be used for validation purposes can be questioned, although it seems reasonable. WBMOD cannot provide time series of perturbed signals.

- GISM [23–25] consists of the NeQuick [26] ionosphere model plus the multiple phase screen (MPS) propagation models to calculate the effects that density irregularities produce in the propagation of electromagnetic waves. It can produce complex time series, and it has been used for most studies assessing ionospheric impact on EGNOS and GALILEO missions [27–30]. GISM is a very powerful model handling arbitrary transmitter and receiver locations, so the incidence angle with respect to the ionosphere layers and to the magnetic field vector orientation can be arbitrary, it can either cross the whole ionosphere or just a part of it. GISM's outputs are the mean effects (total electron content (TEC), distance, phase, angular bend, and Faraday rotation), and the fluctuations characterized by the scintillation parameters (S_4, σ_ϕ, p).

Its main limitations lie in the fact that the scintillation intensity directly depends on the variance of the electron density at any point and any time in space, which is defined by a constant ratio with respect to the ionosphere electron density mean value provided by NeQuick 2, and that only a mean scintillation level is estimated, without statistical variability. GISM is 2D model, and it does not take into account anisotropies. It is a reliable model for equatorial regions, but it is not usable at high latitudes, as the

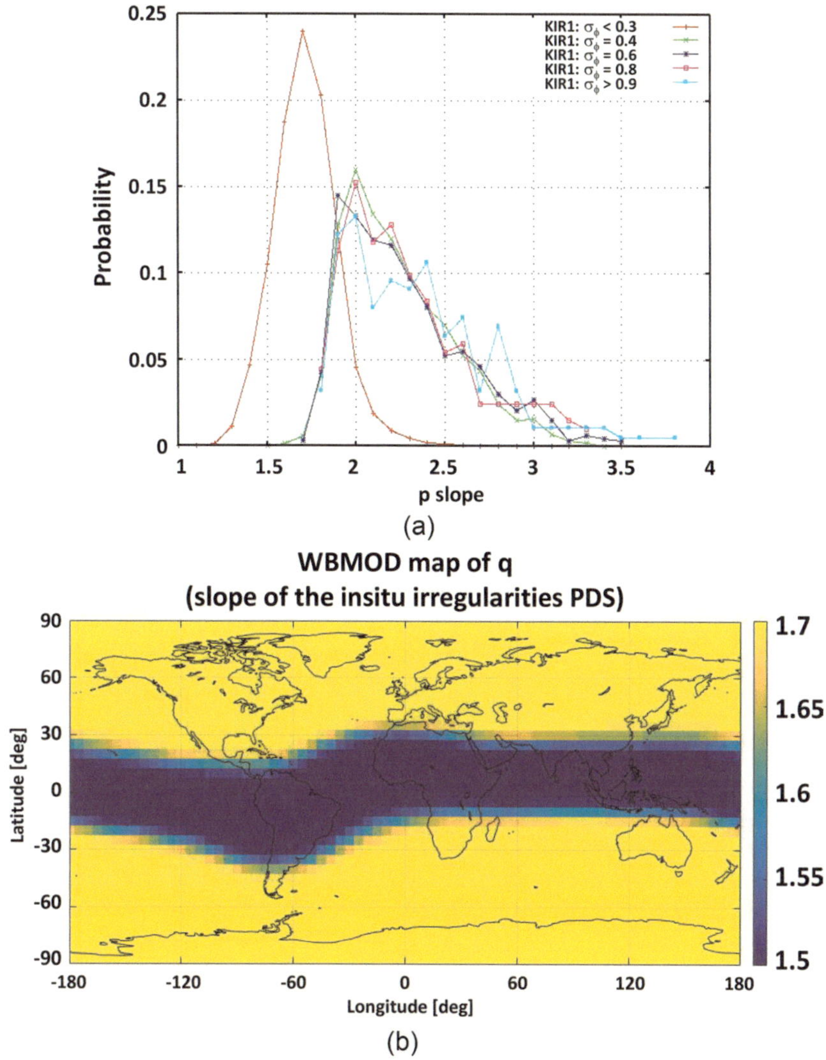

Figure 2.
(a) Measured PDF of the phase spectra slope (p) measured at KIR1 receiver (Kiruna, Sweeden) for different levels of scintillation (σ_ϕ: [0, 0.3], (0.3, 0.5], (0.5, 0.7], (0.7, 0.9], (0.9, ∞), legend indicates central value of the interval). (b) Map of q from WBMOD (p q + 1). In GISM p = 3.

sensitivity to the magnetic activity (e.g., the geomagnetic K_p index[2]) is not accounted for (only in the TEC values from NeQuick), and its forecasting performances are limited as it is driven only by the solar flux number F10.7, which is a daily parameter.

- STIPEE [31] proposes a 2D or 3D formulation of the propagation modeling, based on the parabolic wave equation associated with multiple phase screen model (PWE-MPS) or on the Rytov approximation. The medium is described by the Shkarofsky spectrum [32], then anisotropy and medium drift velocity are taken into account. It is reliable for polar and equatorial regions, and it is valid from

[2] The geomagnetic K_p index can be obtained, for example, from https://kp.gfz-potsdam.de/en/.

weak up to strong scintillation if PWE-MPS resolution is used. STIPEE can be used in conjunction with WBMOD (providing C_kL, drift velocity, slope, and anisotropy) to produce time series, or input parameters can be given by the user (electron density variance and ionospheric spectrum parameters, such as drift velocity, slope and anisotropy). At present, a new prediction model has been derived called HAPEE (high latitude scintillation positioning error estimator) and validated [33]. A merger of STIPEE and HAPEE is under construction within a CNES project.

- SCIONAV [34] is a generic model to evaluate the impact of ionospheric disturbances at low and high latitudes, based on physically-based models for low- and high-frequency fluctuations. SCIONAV uses the total electron content (TEC) parametric model from IRI 2016 [35] or NeQuick [23] as a climatologic background, plus a stochastic variability term computed as a mean value plus a uniform random variable computed from LUTs (Look Up Tables) as a function of the year, month, local time, and latitude. These LUTs have been derived by the UPC/gAGE research group [36] after a comprehensive analysis of the VTEC Global Ionospheric Maps ("Final Products"), published by the International GNSS Service (IGS), from 2001 to 2015 (both included) over a $20° \times 20°$ grid. On top of the background and stochastic TEC, based on the statistics of the equatorial plasma bubbles or EPBs observed during the years 2002–2014, an analytical model developed by the Observatori de l'Ebre [37] accounts for the slowly moving ionospheric depletions and bubbles (EPBs) in equatorial regions (low-frequency TEC fluctuations).

The high-frequency fluctuations are modeled at high latitudes ($\sigma_\phi \neq 0, S_4 \approx 0$) using a model developed by the UPC/gAGE research group in which the phase scintillation is proportional to the *ROTI* (Rate Of TEC Index) parameter computed using 1 Hz data [38, 39], and the mapping function (M):

$$\sigma_\phi = \alpha \cdot M \cdot ROTI. \tag{4}$$

The proportionality constant is α = 0.0555 rad, and the *ROTI* is sum of three terms:

$$ROTI = ROTI_0 + \Delta ROTI + ROTI_{\text{off}} \tag{5}$$

where $ROTI_0$ is proportional to the $AATR$ (Along Arc TEC Rate) [40], $\Delta ROTI$ is a uniform random variable, and $ROTI_{\text{off}} \approx 0.4$ TECUs/min, is an offset term.

At equatorial regions, GISM is used to obtain S_4 and σ_ϕ, and the scintillation enhancement due the presence of EPBs is modeled as an increase of the effective S_4 parameter by a factor $1 + C \cdot |\Delta STEC_{EPBs}(t)|$), where $\Delta STEC_{EPBs}$ is the change in the Slant TEC due to the EPB. Finally, the Cornell Model [41] is used to generate the time series associated to the fast diffractive scintillation.

To summarize, on one side, WBMOD and GISM are both theoretical models calibrated with data on the global morphology of scintillation activity derived from combining measurement data from the VHF and L-band links, but not GNSS data. STIPEE and SCIONAV, being based on WBMOD and GISM inherit the same intrinsic limitations. On the other side, the S_4 model derived from the GNSS radio occultation data from the FormoSat-3/COSMIC mission [14, 15] is a promising empirical model because it includes homogeneous data obtained at global scale, but—

to the authors' knowledge—it still needs to be validated and calibrated using ground-based GPS S_4.

In this study, the goodness of these models is assessed by comparing their outputs to the scintillation parameters measured from a network of GNSS monitoring ground stations. Additionally, these measurements have been used to fine-tune the FormoSat-3/COSMIC S_4 empirical model (or simply COSMIC model in what follows), which can then be used to improve other models in which intensity scintillation is not properly modeled.

2. Methodology and data sets

Performing a validation/verification in the sense of matching model results with experimental data is a necessary proof that the models are correctly developed and implemented, although it might still not be a sufficient proof because experimental data may not be fully representative or suffer from systematic or random errors that would make the model not according to reality, but only according to the reference empirical data set used for model validation. In order to avoid this problem, several steps will be followed, from the simplest case and easier-to-test conditions to increasingly challenging iterations.

Data sets are selected to be as representative as possible according to the following procedure. First, the data will be gathered into four representative regions: Polar caps or PLC (magnetic latitudes greater than 80°), high latitudes or HLT (magnetic latitudes between 70° and 80°), Europe or EUR (excluding the high-latitude regions), and Low/equatorial latitudes or LEQ (±30° around magnetic equator). Second, for each region the data are further divided into severity levels of the observed ionospheric activity: Type 1 corresponding to very high/extreme activity, Type 2 corresponding to moderate to high activity, and Type 3 corresponding to undisturbed ionosphere to low activity. In order to select data for Type 1 scenarios, the events producing the largest AATR values during a full solar cycle period from 2006 to 2016 have been selected. To this end, AATR values were calculated by UPC/gAGE for a network of more than 100 permanent ground stations worldwide distributed. For Type 2 and 3 scenarios, the AATR values were considered for time periods around maximum and minimum solar activity within the interval 2006–2016. In particular, for maximum solar activity, we have focused on the months with average solar flux index F10.7 greater than nearly 150 sfu (solar flux units), which correspond to the periods from November 2013 to April 2014 and from September to December 2014. For minimum solar activity, we have focused on the months with average solar flux index F10.7 smaller than 80 sfu, which are found from February 2007 to December 2009.

Data are provided by different types of ground receivers belonging to two different networks: IGS high-rate geodetic receivers operating at 1-Hz sampling frequency (measurements at intervals of 1 second) and ionospheric scintillation monitor receivers (ISMR) from ESA-MONITOR network [42] providing measurements at 50-Hz sampling frequency (intervals of 0.02 seconds). Part of the data collected from ISMR has been used to verify the reliability of data from 1-Hz geodetic receivers, and another part has been used for model validation, particularly for p-slope (slope of the phase scintillation spectrum) and for S_4 in the equatorial region in South America. Details of which type of data is provided by receivers from the different networks can be found in [38, 39]. Finally, the list of proposed scenarios for model validation based on GNSS data is presented in **Table 1**.

ID label of the scenario	Time period				Scenario description	
	Year	Day of the Year	LT (h) interval	UT (h) interval	Solar activity	Comments
Type 1 Scenarios						
Polar Cap and High-Latitude Regions						
GNSS-Typ1-HLT-1	2015	76	—	[15, 19]	High	St. Patrick's storm
GNSS- Typ1-HLT-2	2014	58	—	[19, 24]	High near maximum	Minor storm but frequent high AATR
GNSS-Typ1-HLT-3	2015	250	—	[15, 24]	High	G2 geomag. Storm
GNSS-Typ1-HLT-4	2015	252	—	[02, 08] & [13, 16]	High	G2 geomag. Storm
Europe Region						
GNSS- Typ1-EUR-1	2015	76	—	[15, 19]	High	St. Patrick's storm
GNSS- Typ1-EUR-2	2014	58	—	[12, 17]	High near maximum	Minor storm but frequent high AATR
Low-Equatorial Regions						
GNSS- Typ1-LEQ-1	2014	351	[19, 23]	—	High	Highest AATR in current solar cycle
GNSS- Typ1-LEQ-2	2014	314	[19, 24]	—	High	Activity most equatorial stations
GNSS-Typ1-LEQ-3	2015	298 299	[19, 24]	—	High	Highest activity in days of Set 2
Type 2 Scenarios						
Polar Cap and High-Latitude Regions						
GNSS- Typ2-HLT-1	2015	291	—	[06, 19]	High	Minor storm. High activity most stations
GNSS- Typ2-HLT-2	2014	62	—	[04, 13]	Near Maximum	High Ionospheric activity
GNSS- Typ2-HLT-3	2007	91	—	[02, 09]	Low near Minimum	Minor storm.
Europe Region						
GNSS- Typ2-EUR-1	2014	87	—	[12, 17]	Near Maximum	High Ionospheric activity
GNSS- Typ2-EUR-2	2007	155	[16, 20]	—	Low near Minimum	Moderate Ionospheric activity
GNSS-Typ2-EUR-3	2015	116	[09, 20]	—	High	Highest activity in the region for days of Set 2
GNSS-Typ2-EUR-4	2015	291	[09, 16]	—	High	Minor storm effects in the region
Low-Equatorial Regions						
GNSS- Typ2-LEQ-1	2014	87	[19, 24]	—	Near Maximum	High Ionospheric activity
GNSS- Typ2-LEQ-2	2007	80	[19, 24]	—	Low near Minimum	High Ionospheric activity

ID label of the scenario	Time period				Scenario description	
	Year	**Day of the Year**	**LT (h) interval**	**UT (h) interval**	**Solar activity**	**Comments**
GNSS-Typ2-LEQ-3	2015	291	[19, 24]	—	High	Only stations in South America & Canary Island
GNSS-Typ2-LEQ-3	2015	115	[19, 24]	—	High	Stations: mal2, dgar, pimo, LOM2, DAK2
Type 3 Scenarios						
Polar Cap and High-Latitude Regions						
GNSS- Typ3-HLT-1	2014	257	—	[15, 21]	Near Maximum	Low Ionospheric activity
GNSS- Typ3-HLT-2	2007	181	—	[10, 15]	Near Maximum	Low Ionospheric activity
GNSS-Typ3-HLT-3	2015	299	—	[05, 20]	High	Low/Quiet ionospheric activity
Europe Region						
GNSS- Typ3-EUR-1	2015	252	[00,09] & [19, 24]	—	High	Quiet Ionosphere nearly all day
GNSS- Typ3-EUR-2	2014	20	[01, 05]	—	Near Maximum	Quiet Ionosphere
GNSS- Typ3-EUR-3	2007	181	—	[10, 15]	Near Maximum	Low Ionospheric activity
Low-Equatorial Regions						
GNSS- Typ3-LEQ-1	2015	250–251	[01, 09]	—	High	Quiet Ionosphere, scintillation activity very scarce, the lowest for days in Set 2
GNSS- Typ3-LEQ-2	2014	20	[01, 05]	—	Near Maximum	Quiet Ionosphere
GNSS- Typ3-LEQ-3	2007	181	—	[10, 15]	Near Maximum	Low Ionospheric activity

Table 1.
Consolidated list of GNSS test scenarios.

In the case of the SCIONAV model, a different data set has been used to extract the parameters needed to properly tune the model. Then, these parameters have been organized in different look-up tables (LUTs) by date, latitude, and any other variable that can potentially influence the value of the corresponding parameter (e.g., solar activity). Subsequently, in order to test the model performance, the data set described in **Table 1** has been chosen approximately at the same locations and times as the data set used to tune the parameters of the SCIONAV model. Specifically, the GNSS test scenarios in **Table 1** are within a range of similar conditions as the data used for model parameter tuning, that is, similar pierce-point locations (elevation angle, magnetic latitude, azimuth direction with regard to solar zenith angle, and magnetic field lines), and similar geomagnetic activity, time-of-day (day/night, close to sunset or not), or conditions depending both of local time and season of the year. In this way, the observed scintillation parameters (S_4, σ_ϕ, ROTI and power spectra slope), and the ionospheric activity index (AATR) can be consistently compared to the scintillation parameters predicted by the original models in similar ionospheric conditions. Note that, the AATR is a widely used indicator of ionospheric activity linked to the local TEC fluctuations observed from GNSS measurements taken by permanent ground receivers. For this reason, it has been used to select the different types of scenarios proposed in the present study according to the level of activity derived by using the AATR thresholds established by [40] for the regions with different ionospheric activity that essentially correspond to the different regions considered in the present work. The AATR has been demonstrated to be correlated with space weather conditions and can be used to monitor planetary storm conditions [43].

Finally, the list of proposed scenarios for model validation based on GNSS data is presented in **Table 1**.

Note that, for every region, the data are provided by different ground stations at different locations. When the ionospheric activity occurs at a planetary scale, as in the case of some space weather events or geomagnetic storms, the universal time is used to define the data set of the given scenario. Instead, when the ionospheric activity is linked to local times, as in the case of low-latitude scintillation, the scenario is better defined using the local time range, which is relevant at all low-latitude locations.

3. Results

In the next set of plots, the general behavior of the models is compared against the data. Although the plots are divided into geographical regions, they are not subdivided into scintillation categories. Large differences between different low-latitude regions (e.g., between America and Asia/Africa) were not observed in the reference data sets used for model testing, nor in the corresponding model predictions. Therefore, all data from low-latitude regions have been joined in a single set to perform the model analysis.

3.1 S_4 results

The S_4 model results are only available for GISM and WBMOD, as the SCIONAV model uses the same S_4 as GISM. Results are summarized in **Figures 3–5. Figure 3** shows the S_4 PDF in percentage (%) from the ground measurements, from the GISM model, and from the WBMOD + STIPEE model, and per region: equatorial low latitudes, Europe, high latitudes, and polar caps. In the LEQ region, S_4 values have low probability, but they exist, although the scale of the figure does not help to appreciate that. Instead, at PLC and HLT large values of S_4 are not expected since scintillation is mostly refractive and affects the carrier phase but not the signal intensity. Similarly, **Figure 4** shows the density plots of S_4 vs. the *ROTI*, and **Figure 5** shows the density plots of S_4 vs. the local time (*LT*). Note that, in the case of ISMR receivers operating at 50 Hz, the values of σ_ϕ and p are directly outputs of the receivers, while the ROTI is straightforwardly derived from the TEC variations output by the receivers. In the case of geodetic receivers from the IGS network, operating at 1 Hz sampling frequency, the parameters were calculated after processing the carrier phase measurements from the receivers following the Geodetic Detrending methodology as described in Refs. [38, 39].

It has been observed in real data that S_4 is correlated with ROTI at low latitudes [44, 45]. However, in **Figure 3** model predictions cannot probably reproduce the S_4

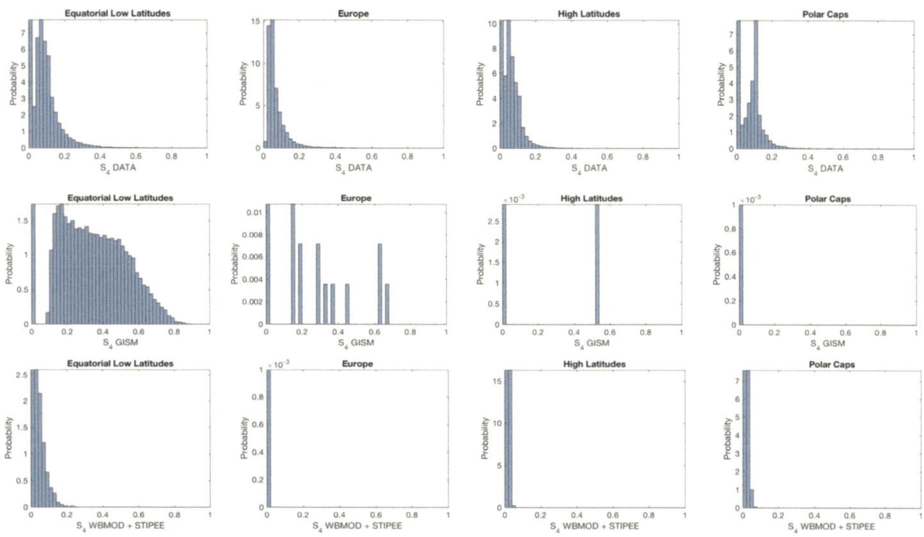

Figure 3.
PDF [in %] of: (first row) S_4 *from the ground measurements (scale normalized to maximum value because of large variations between regions and models), (second row) from the GISM model (SCIONAV has the same output), and (third row) from the WBMOD + STIPEE model, per region.*

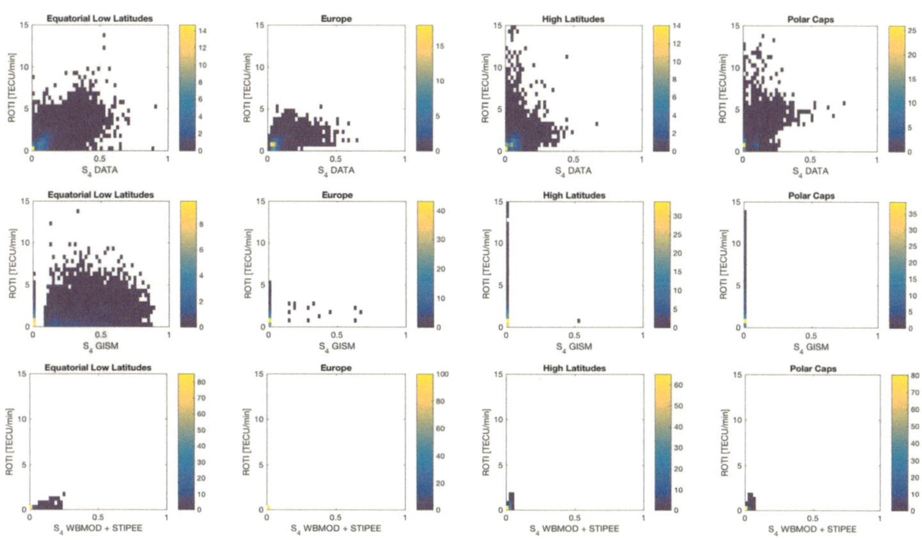

Figure 4.
Density plots of: (first row) S_4 *vs.* ROTI *from the ground measurements, (second row) from the GISM model (SCIONAV has the same output), and (third row) from the WBMOD + STIPEE model, per region. Colorbar indicates percentage.*

data with sufficient accuracy, which is probably responsible of the bad results concerning the S_4-ROTI correlation. As it can be seen, S_4 PDFs from WBMOD +STIPEE are clearly better than those from GISM. However, the dependence with respect to *ROTI* and *LT* seems to be better captured by GISM, notably at equatorial regions, not so good at mid-latitudes, and —as expected— not at all at polar regions since it is known that GISM is not modeling properly S_4 at high latitudes.

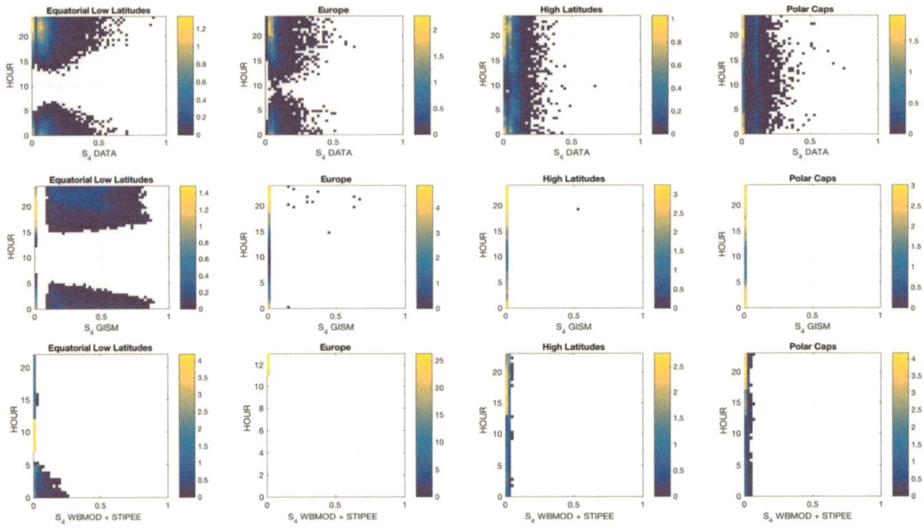

Figure 5.
Density plots of: (first row) S_4 *vs. LT from the ground measurements, (second row) from the GISM model (SCIONAV has the same output), and (third row) from the WBMOD + STIPEE model, per region. Colorbar indicates percentage.*

3.2 σ_ϕ results

The σ_ϕ model results are available for all models, as SCIONAV includes a specific model for the phase scintillation [G. Gonzalez-Casado and J.M. Juan-Zornoza, internal communication, 2016].

Figure 6 shows the σ_ϕ [in radians] PDF in percentage (%) from the ground measurements, from the GISM model, from the SCIONAV model, and from the WBMOD + STIPEE model, and per region: equatorial low latitudes, Europe, high latitudes, and polar caps. Similarly, **Figure 7** shows the density plots of σ_ϕ vs. the *ROTI*, and **Figure 8** shows the density plots of σ_ϕ vs. the local time (*LT*).

As it can be seen, σ_ϕ PDFs from SCIONAV are qualitatively better than the other models, WBMOD+STIPEE model being slightly worse than SCIONAV model, as it is a bit wider (i.e., predicts slightly larger σ_ϕ values), and GISM clearly overestimates σ_ϕ, notably at equatorial low latitudes and Europe regions. The dependence with respect to *ROTI* and *LT* seems better captured by the SCIONAV model at all latitudes, although some improvements may still be needed at high latitudes to increase the correlation. For WBMOD+STIPEE the correlation is too high, and the dynamic range is too small.

3.3 p-slope results

The p-slope model results are available for SCIONAV and WBMOD+STIPEE models. GISM assumes a constant value of $p = 3$ [24]. During the analysis of this data, it became apparent that there may be some discrepancies in the definition of the p-slope. On one hand, WBMOD values agree with Beniguel et al. findings [42]. On the other hand, the LUTs used for the p-slope by the SCIONAV model were calculated using values delivered by 50 Hz ISMR from the MONITOR network. In **Figure 2a** typical p-slope distributions depending on scintillation severity are shown. **Figure 2b**

DOI: http://dx.doi.org/10.5772/intechopen.1001077

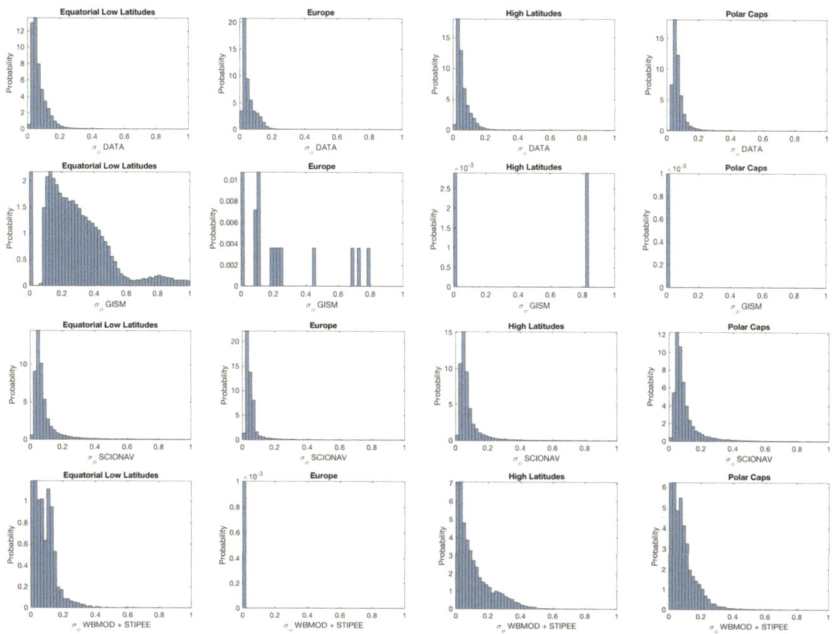

Figure 6.
PDF [in %] of: (first row) σ_ϕ *PDF [in %] from the ground measurements (scale normalized to maximum value because of large variations between regions and models), (second row) from the GISM model, (third row) from the SCIONAV model, and (fourth row) from the WBMOD + STIPEE model, per region.*

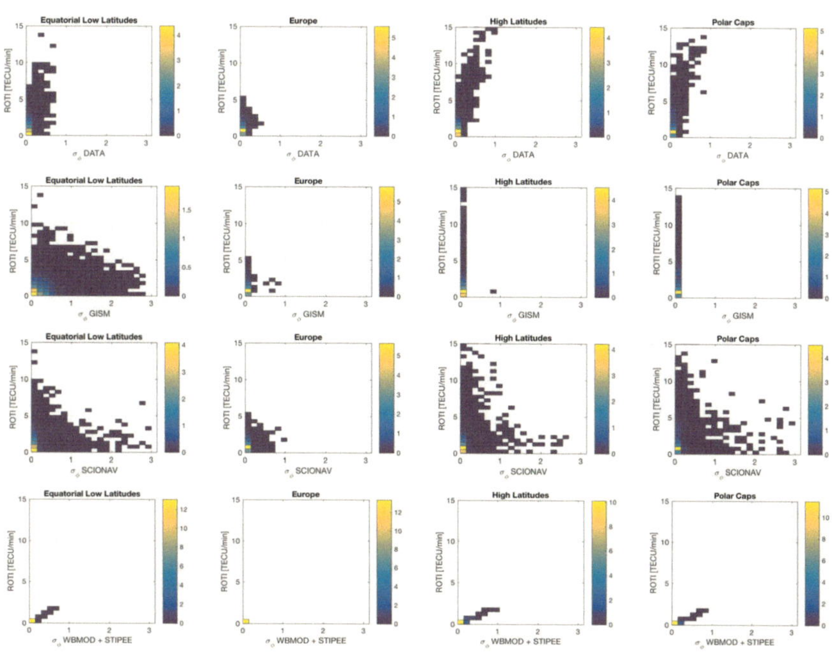

Figure 7.
Density plots of: (first row) σ_ϕ *vs.* ROTI *from the ground measurements, (second row) from the GISM model, (third row) from the SCIONAV model, and (fourth row) from the WBMOD + STIPEE model, per region. Colorbar indicates percentage.*

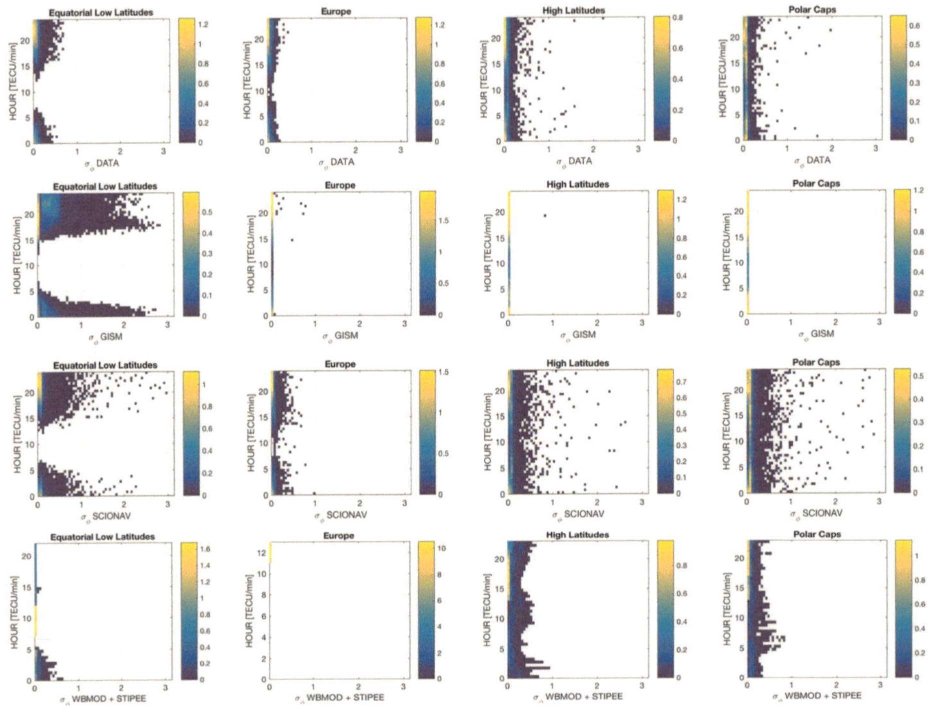

Figure 8.
Density plots of: (first row) σ_ϕ vs. LT from the ground measurements, (second row) from the GISM model, (third row) from the SCIONAV model, and (fourth row) from the WBMOD + STIPEE model, per region. Colorbar indicates percentage.

shows the maps of the q-slope obtained from WBMOD (WBMOD outputs q, being $p \sim q + 1$) [22]. As it can be seen, it basically has two values: $q = 1.5$ at equatorial low-latitude regions and $q = 1.7$ at mid-to-polar latitudes with a narrow transition zone. Values of p between $p = 2.5$ and $p = 2.7$ should then be expected according to WBMOD, but even for high scintillation events, the p-slope values rarely go above 2.5. ISMR observations used to build the distributions shown in **Figure 2a** show that the probability of $p > 2.5$ is just around 15–20% for periods of scintillation activity ($\sigma_\phi > 0.3$).

Figure 9 shows the p-slope PDF in percentage from the ground measurements, from the GISM model, from SCIONAV, and from the WBMOD + STIPEE model, and per region: equatorial low latitudes, Europe, high latitudes, and polar caps. Similarly, **Figure 10** shows the density plots of p-slope vs. the *ROTI*, and **Figure 11** shows the density plots of p-slope vs. the *LT*.

Neither GISM (constant p-value equal to 3, not shown), nor WBMOD (fixed value depending on latitude $q \sim p\text{-}1 \sim 1.5$ or 1.7, with narrow transition, **Figure 2b**), nor SCIONAV exhibit the natural variability, and the variability has to be increased in SCIONAV. When looking at **Figures 10** and **11**, it can be observed that the p - slope values from SCIONAV exhibit an artificial binomial behavior, which can be attributed to the way the scenarios were selected: groups of low-phase scintillation $p \sim 1.7$, and others of high phase scintillation $p \sim 2.3$ (see **Figure 2a**). Apart from that, the SCIONAV model exhibits the correct dependence with *ROTI*, including the dynamic range and so with *LT*, which does not seem to be the case of WBMOD+STIPEE.

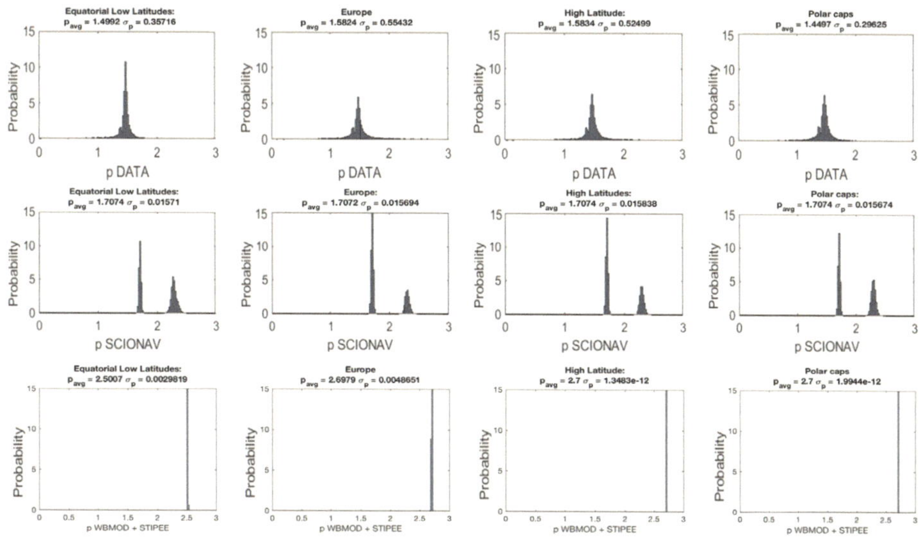

Figure 9.
PDF [in %] of: (first row) p-slope *PDF from the ground measurements, (second row) from the SCIONAV model, and (third row) from the WBMOD + STIPEE model, per region.*

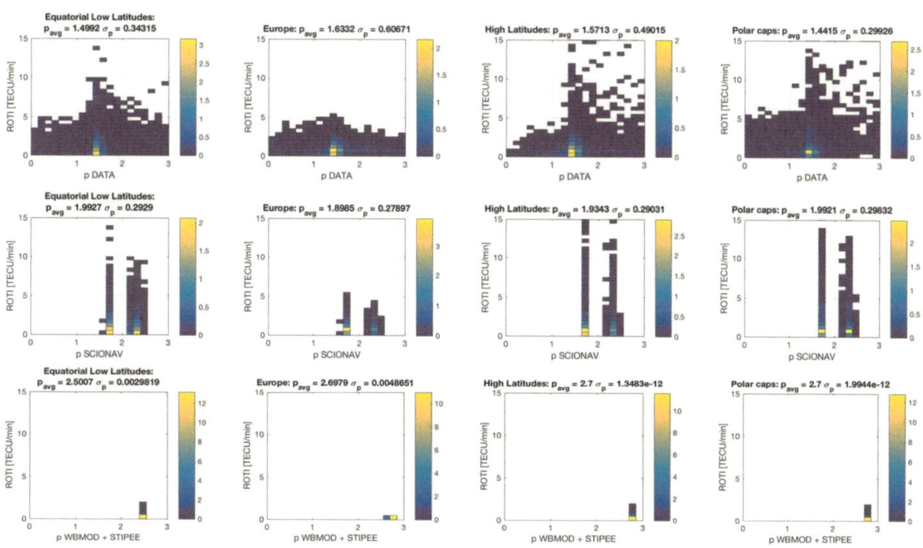

Figure 10.
Density plots of: (first row) p-slope *vs.* ROTI *from the ground measurements, (second row) from the SCIONAV model, and (third row) from the WBMOD + STIPEE model, per region. Colorbar indicates percentage.*

3.4 Model performance summary

Quantitative results of the previous sections are summarized in **Tables 2–4** in terms of the mean and unbiased root mean squared error (uRMSE). A word of caution is given to **Table 4** because of the difference in one unit in the definition of the slope (p in all models, and q in WBMOD).

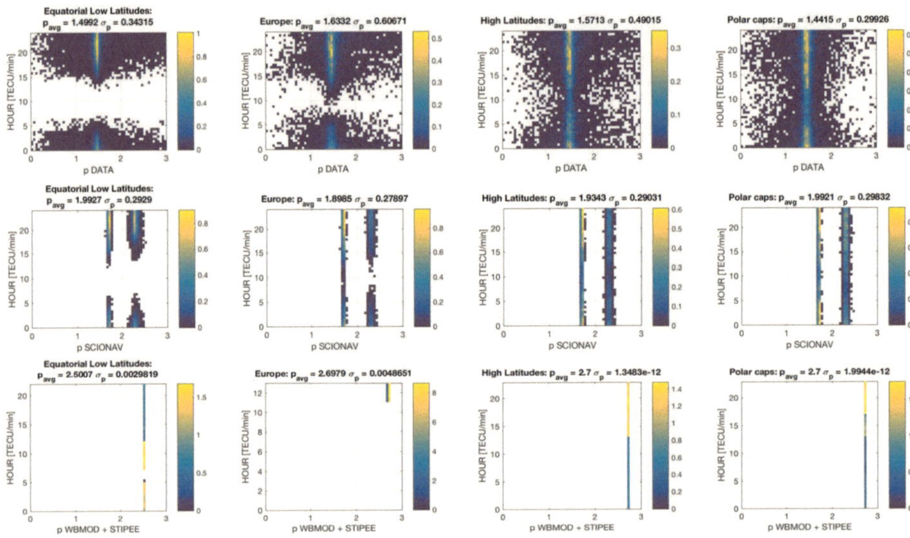

Figure 11.
Density plots of: (first row) p-slope vs. LT from the ground measurements, (second row) from the SCIONAV model, and (third row) from the WBMOD + STIPEE model, per region. Colorbar indicates percentage.

The following conclusions on the models were drawn:

- S_4 modeling analysis:
  - GISM does not seem to have a good S_4 model when compared against the data or against WBMOD. It provides artificially high values of S_4, and it is also known to lack a proper model at high latitudes (results in purple in **Table 2**). It is the same for SCIONAV, which is based on GISM.

		SCIONAV		GISM		WBMOD + STIPEE	
		Mean	**uRMS**	**Mean**	**uRMS**	**Mean**	**uRMS**
Low/equatorial latitudes (LEQ)	All types	**0.1744**	**0.2166**	**0.1744**	**0.2166**	**−0.0922**	**0.0783**
	Type 1	0.1510	0.2023	0.1510	0.2023	−0.0691	0.0983
	Type 2	0.1957	0.2380	0.1957	0.2380	−0.1771	0.1079
	Type 3	0.2333	0.2237	0.2333	0.2237	−0.0209	0.0230
Europe (EUR)	All types	**−0.0679**	**0.0553**	**−0.0679**	**0.0553**	**−0.0385**	**0.0136**
	Type 1	—	—	—	—	—	—
	Type 2	−0.0681	0.0557	−0.0681	0.0557	−0.0385	0.0136
	Type 3	—	—	—	—	—	—
High latitudes (HLT)	All types	**−0.0575**	**0.0516**	**−0.0575**	**0.0516**	**−0.0210**	**0.0263**
	Type 1	−0.0585	0.0531	−0.0585	0.0531	−0.0193	0.0269
	Type 2	—	—	—	—	—	—
	Type 3	−0.0557	0.0507	−0.0557	0.0507	−0.0234	0.0251

		SCIONAV		GISM		WBMOD + STIPEE	
		Mean	**uRMS**	**Mean**	**uRMS**	**Mean**	**uRMS**
Polar caps (PLC)	All types	−0.0502	0.0634	−0.0502	0.0634	−0.0922	0.0412
	Type 1	−0.0659	0.0637	−0.0659	0.0637	−0.0246	0.0446
	Type 2	—	—	—	—	—	—
	Type 3	−0.0726	0.0732	−0.0726	0.0732	−0.0283	0.0402

Bold type: All types of scenarios; green color: Best results, purple color: Model not applicable

Table 2.
Mean (left) and uRMSE (right) error for S_4. Note: GISM results at high latitudes and polar caps are not meaningful.

		SCIONAV		GISM		WBMOD + STIPEE	
		Mean	**uRMS**	**Mean**	**uRMS**	**Mean**	**uRMS**
Low/equatorial latitudes (LEQ)	All types	0.0182	**0.1356**	**0.2133**	**0.3410**	**−0.0724**	0.0653
	Type 1	0.0484	0.2190	0.1623	0.2437	−0.0434	0.0862
	Type 2	0.0133	0.1085	0.4540	0.5875	−0.1345	0.1051
	Type 3	0.0076	0.0808	0.2907	0.4184	−0.0333	0.0198
Europe (EUR)	All types	−0.0055	**0.0629**	**−0.0541**	**0.0394**	**−0.0479**	0.0280
	Type 1	—	—	—	—	—	—
	Type 2	−0.0033	0.0610	−0.0522	0.0375	−0.0479	0.0280
	Type 3	—	—	—	—	—	—
High latitudes (HLT)	All types	0.0206	**0.1392**	−0.0625	0.0643	**0.0378**	**0.1196**
	Type 1	0.0563	0.2857	−0.0937	0.1164	0.0730	0.1418
	Type 2	—	—	—	—	—	—
	Type 3	0.0082	0.0663	−0.0502	0.0316	−0.0145	0.0330
Polar caps (PLC)	All types	**0.0402**	**0.1798**	−0.0686	0.0542	0.0107	0.0876
	Type 1	0.0599	0.2274	−0.0791	0.0742	0.0458	0.1024
	Type 2	—	—	—	—	—	—
	Type 3	0.0284	0.1151	−0.0650	0.0287	−0.0244	0.0201

Bold type: All types of scenarios; green color: Best results, purple color: Model not applicable.

Table 3.
Mean (left) and uRMSE (right) error for σ_ϕ.

		SCIONAV		WBMOD + STIPEE	
		Mean	**uRMS**	**Mean**	**uRMS**
Low/equatorial latitudes (LEQ)	All types	0.4936	**0.4538**	**0.9817**	0.3364
	Type 1	0.5188	0.3690	0.9776	0.2787
	Type 2	0.4907	0.3922	0.9620	0.2778
	Type 3	0.4394	0.5271	0.9912	0.386

		SCIONAV		WBMOD + STIPEE	
		Mean	**uRMS**	**Mean**	**uRMS**
Europe (EUR)	All types	0.2662	**0.6361**	**1.1162**	0.4270
	Type 1	—	—	—	—
	Type 2	0.2904	0.5937	1.1162	0.4270
	Type 3	—	—	—	—
High latitudes (HLT)	All types	0.3639	**0.5773**	**1.0521**	0.4992
	Type 1	0.2482	0.5685	0.9713	0.5386
	Type 2	—	—	—	—
	Type 3	0.4257	0.4494	1.1721	0.4054
Polar caps (PLC)	All types	0.5507	**0.4285**	**1.239**	0.2985
	Type 1	0.6402	0.4071	1.202	0.3432
	Type 2	—	—	—	—
	Type 3	0.6884	0.3245	1.315	0.2517

Bold type: all types of scenarios; green color: best results.

Table 4.
Mean (left) and uRMSE (right) error for p*-slope.*

- o WBMOD+STIPEE has a more realistic distribution, even if it produces values slightly lower than reality.
- o While it has been observed in real data that S_4 is correlated with ROTI at low latitudes [40, 42], this correlation is not captured by any model. This is a motivation to investigate the performance of the COSMIC model in order to improve the S_4 modeling.
- o GISM and WBMOD models account properly for the dependence with LT.
- o In this study, SCIONAV model has the bubbles and depletions code deactivated, so the estimated S_4 values would be slightly higher.

• σ_ϕ modeling analysis:

- o GISM models artificially high values of $\sigma_{\phi.}$
- o SCIONAV model exhibits a very good agreement with data in terms of mean and standard deviation, for all regions, types of events, and LT.
- o WBMOD model is very good, but in some cases, it does not reproduce so well the dispersion with LT, and it has lower error dispersion across regions.

• p-slope modeling analysis:

- o GISM uses a default value of p = 3.

- WBMOD uses a fixed value depending on latitude (q = 1.5 at equatorial low latitudes, or q = 1.7 at mid-to-polar latitudes; so $p \sim q + 1 = 2.5–2.7$).

- SCIONAV modeled p-slope PDF depends on σ_ϕ (**Figure 2a**). The bimodality presented could be caused by the simplicity of the modeling based on the PDFs presented in **Figure 2a**, and/or by the selection of the data in extreme cases.

- Additionally, estimated values of p from real data are biased towards lower values because of the limitations of the IGS 1 Hz data, which does not sample the full spectra of phase fluctuations.

- The p-slope is not the best-suited parameter to assess the performance analysis of the models considered in the present study.

4. Discussion and way forward

4.1 Summary of the experimental results

So far, the assessment of model performance was done using GNSS data (i.e., L-band). A fundamental point to keep in mind is the difference modeling philosophy in the four models.

Whereas GISM and SCIONAV are providing a mean level of scintillation along each link, WBMOD and STIPEE are considering the statistics of occurrence of an event, or a scintillation index value for a given percentile. In order to compare the different models, WBMOD/STIPEE results have been estimated at a percentile 50.

One additional important point is that the models' inputs are different. Along Arc TEC Rate (*AATR*) scintillation index is used as input for SCIONAV to characterize the phase scintillation, whereas WBMOD is using the *Kp* index, which is not a scintillation index, it is a geomagnetic index derived from the standardized *K*-index of 13 magnetic observatories, and it is designed to measure the solar particle radiation by its magnetic effects. In this way the approaches of the models are different, WBMOD being a blind model considering ionosphere scintillation input but considering magnetic conditions instead.

GISM results have been assumed as a mean level of scintillation. This assumption has been taken to compare the model with measurements, but it can be discussed. The problem remains in the definition, and in the derivation of the climatologic part of the models that have not been well described by the models. WBMOD is based on a statistical approach, considering that for a given ionospheric conditions (date, local time, solar exposure, latitude, and magnetic conditions) the scintillation intensity may change with time. An illustration is displayed in **Figure 12**, representing ISMR S_4 data measured in Parepare, Indonesia, and the corresponding WBMOD prediction at the ninetieth percentile. Four consecutive days have been chosen in September 1999. These plots demonstrate the "patchy" nature of scintillation in contrast to WBMOD, which produces smooth predictions of scintillation for a given percentile. WBMOD, such as all other models proposed in this study, is not able to reproduce the night-to-night variability of scintillation activity. This variable behavior of the measurements indicates that a statistical approach seems to be required to model scintillation.

Medium anisotropy influence is of high importance for SAR modeling, and it has also an impact on GNSS links. WBMOD is proposing a parametrization of anisotropy

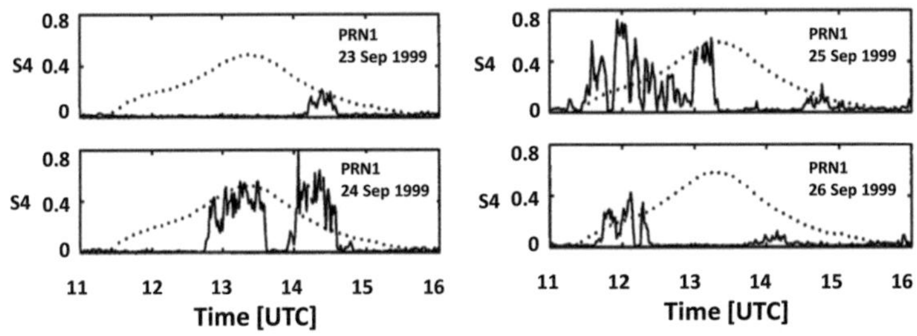

Figure 12.
Comparison between ISMR S_4 (solid line) data and WBMOD predictions at 90 percentile (dotted line) on five separate days in September 1999 for GPS satellite Pseudo-Random code Number (PRN) 1 observed at Parepare, Indonesia (adapted from [19]).

in terms of the along-field: (a) across field and (b) axial ratios, with values from ~10–50 and ~ 1–4, respectively. These variables exhibit a smooth evolution with the local solar time (LST) and K_p, but no dependence with the Day of the Year (DoY), the twelve-month smoothed relative sunspot number (R12), or the percentile. However, this remains hard to validate.

In a conclusion, as it has been shown, while the phase scintillations are reasonably well characterized by both SCIONAV and WBMOD+STIPEE models (see **Table 3**), large errors occur in S_4 in GISM and SCIONAV, which is based on GISM for S_4 (see **Table 2**). Hence, any potential model improvement in S_4 predictions should be very relevant and will help to improve the overall accuracy.

4.2 S_4 model improvement

In the following paragraphs, an attempt to improve the S_4 modeling is described using the semiempirical model developed in [14, 15] to convert the 3D FORMOSAT-3/COSMIC or F3/C $S_{4,\ max}$ (maximum value on each profile) into a 2D latitude and longitude S_4-index map on the ground has been used (hereafter called S_4-index).

The scintillation model can be subdivided into low-latitude and high-latitude parts, which are connected/overlapped between ±45° and ± 65° dip. The model combines a constant low-latitude weight, between −45° and 45° dip latitudes, and a constant high- latitude weight, below −65° and above 65° dip latitude, with linear transitions between those two regions in the middle latitudes in both hemispheres. For each region, the model is composed of the product of four terms [15], the Diurnal one, S_{Diur}, the annual one, S_{An}, the one that models the latitudinal (dip) variations, S_{Dip}, and the one that depends on the solar flux, $S_{PF10.7}$, as shown below:

$$S_4 = S_{Diur}(LT) \cdot S_{An}(DoY, lon) \cdot S_{Dip}(Dip) \cdot S_{PF10.7}(YY, DoY, PF10.7) \cdot k + S_{4,bias}. \tag{6}$$

In (Eq. (6)) the term *YY*, *DoY*, and *LT* are the Year, Day of the Year and local time, is the *PF*107 = (*F*10.7+ *F*10.7*A*)/2, being *F*10.7 the solar flux daily value, and *F*10.7*A* the 81-day running mean value of the solar radio flux at 10.7 cm wavelength, respectively.

Originally, the model used k = 0.78125 and $S_{4,bias}$ = 0.07. However, to achieve consistency with the ground GNSS data set used for model testing (**Table 1**); in the present study, two of the parameters of the model, the calibration constant k and the bias term $S_{4,\text{bias}}$ have been adjusted to minimize the rms error to the values shown in **Table 5**. The remaining parameters and functionalities introduced in [15] have not been modified. Note that biases are negligible, but values of k are much larger than in [15].

Figure 13 shows the performance of the COSMIC model in terms of the S_4 PDF, S_4 error PDF, the density plots of S_4 vs. *ROTI*, and the density plots of S_4 vs. *LT*, per region. Clearly, COSMIC S_4 PDFs seem better than from GISM/SCIONAV, with a noticeable improvement at equatorial regions. ROTI dependence seems also better behaved with COSMIC S_4, except at high latitudes. Finally, *LT* dependence seems also better behaved with COSMIC, at all latitudes. Additionally, the error histograms (second row) are more symmetric and Gaussian-like than for other models (not shown), despite the rms is slightly larger —except in LEQ region—, which is an indication that the model is properly modeling the nature of the statistics of the ionospheric intensity scintillations.

	LEQ	EUR	HLT	PLC
K	7.9902	25.6787	8.9855	9.5047
$S_{4,bias}$	−0.0058	−0.0000	−0.0000	−0.0000

*Values obtained by minimized rms error between experimental data (**Table 1**) and COSMIC model.*

Table 5.
Optimum values of k *and* $S_{4,\text{bias}}$ *for each region: equatorial and low latitudes (LEQ), Europe (EUR), high latitudes (HLT), and polar caps (PLC).*

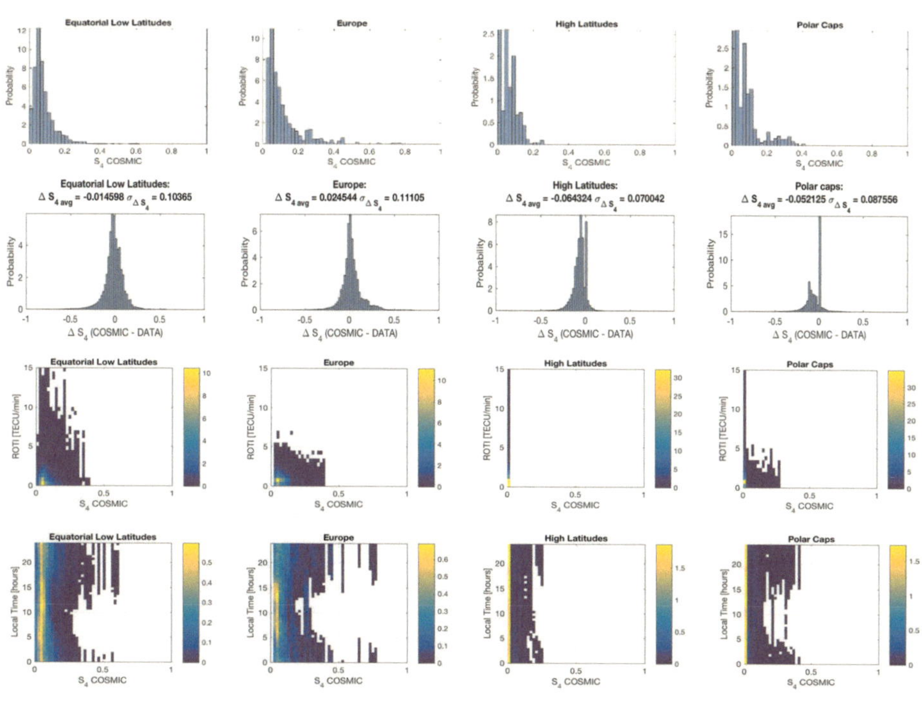

Figure 13.
Performance of COSMIC model: (first row) S_4 *PDF, (second row)* S_4 *error PDF, (third row)* S_4 *vs. ROTI, and (fourth row)* S_4 *vs. LT, per region. Colorbar indicates percentage.*

S_4	SCIONAV/GISM		WBMOD + STIPEE		COSMIC	
	Mean	RMS	Mean	RMS	Mean	RMS
Low/equatorial latitudes (LEQ)	**0.1744**	0.2166	**−0.0922**	0.0783	−0.0146	**0.1037**
Europe (EUR)	−0.0679	**0.0553**	**−0.0385**	0.0136	0.0245	0.1110
High Latitudes (HLT)	−0.0575	0.0516	−0.0210	0.0263	-0.0643	0.0700
Polar Caps (PLC)	−0.0502	0.0634	−0.0922	0.0412	−0.0521	0.0876

Bold type: all types of scenarios; green color: best results, red color: worst results, purple color: model not applicable. Recall that SCIONAV uses GISM S_4 as an input.

Table 6.
Summary of mean and uRMS error for S_4 for all types, regions, and models.

5. Conclusions

This chapter has summarized the outcomes of an ESA study to analyze the performance of several ionospheric scintillation models and to provide recommendations for future improvements. The models analyzed are GISM, SCIONAV, WBMOD, STIPEE, and COSMIC. The analysis has been performed using S_4, σ_ϕ, and p for the first four and, only S_4 for COSMIC, derived from GNSS data from ground stations. Data have been binned by regions: equatorial and low latitudes, Europe and mid-latitudes, high latitudes, and polar caps.

Table 6 summarizes the main parameters of S_4 for all regions, types of scintillation, and models. The GISM model (and SCIONAV, as it uses GISM inside to compute S_4) is not appropriate for high latitudes and polar regions, in terms of S_4. The best results are obtained with a tuned COSMIC model (bias and gain factor) for all regions, except for high latitudes where the WBMOD+STIPEE model outperforms. This is possibly due to the lower data sampling used to derive the COSMIC model.

In terms of σ_ϕ, the best results are obtained by SCIONAV's phase scintillation model, except for polar caps, where WBMOD slightly outperforms. WBMOD outperforms always in terms of the uRMSE.

Finally, in terms of p, SCIONAV model outperforms in terms of the average error, but WBMOD+STIPEE in terms of the uRMSE. Improved modeling of p is still required.

Tuning of the COSMIC S_4 empirical model to the data collected (**Table 1**) has led to the lowest average errors (except for WBMOD at high latitudes). It could be used in conjunction with the SCIONAV σ_ϕ model to better model the ionospheric scintillation behavior.

Future work must be conducted to improve/update the characterization of the patchy behavior of the ionosphere, assess the validity of some approximations of the models, such as the constant slope and the anisotropy … and implement a 4D (3D + time) simulator of the ionospheric scintillation, including ionospheric drifts.

Acknowledgements

This research was funded by the project "Radio Climatology Models of the Ionosphere: Status and Way Forward," ESA/ESTEC, grant number 4000120868/17/NL/AF [https://nebula.esa.int/content/radio-climatology-models-ionosphere-status-and-wa y-forward]. Article processing charges were funded by the project "GENESIS: GNSS Environmental and Societal Missions – Subproject UPC," AEI Grant PID2021-126436OB-C21.

IGS data were available from: https://cddis.nasa.gov/archive/gnss/data/highrate/. MONITOR data were provided by ESA. WBMOD simulations were performed by ONERA; GNSS ground stations data processing was performed by UPC/gAGE, and GISM and SCIONAV simulations were performed by UPC/CommSensLab-IEEC.

Conflict of interest

The authors declare no conflict of interest.

Author details

Adriano Camps[1,2,3]*, Carlos Molina[1], Guillermo González-Casado[4], José Miguel Juan[4], Joël Lemorton[5], Vincent Fabbro[5], Aymeric Mainvis[5], José Barbosa[6] and Raúl Orús-Pérez[7]

1 Department of Signal Theory and Communications, Universitat Politècnica de Catalunya, Barcelona, Spain

2 Institut d'Estudis Espacials de Catalunya/CTE-UPC, Barcelona, Spain

3 United Arab Emirates University, Al Ain, Abu Dhabi, United Arab Emirates

4 Departament de Matemàtiques, Universitat Politècnica de Catalunya, Barcelona, Spain

5 ONERA/DEMR, Université de Toulouse, Toulouse, France

6 RDA-Research and Development in Aerospace GmbH, Zürich, Switzerland

7 ESTEC, Noordwijk, The Netherlands

*Address all correspondence to: adriano.jose.camps@upc.edu

References

[1] Priyadarshi S. A review of ionospheric scintillation models. Surveys in Geophysics. 2015;**36**:295-324

[2] Aarons J. Equatorial scintillations: A review. IEEE Transactions on Antennas and Propagation. 1977;**25**(5): 729-736. DOI: 10.1109/TAP.1977.1141649

[3] Morrissey TN, Shallberg KW, Van Dierendonck AJ, Nicholson MJ. GPS receiver performance characterization under realistic ionospheric phase scintillation environments. Radio Science. 2004;**39**:RS1S20. DOI: 10.1029/2002RS002838

[4] Kintner PM, Humphreys T, Hinks J, GNSS and Scintillation. How to survive the next solar maximum. Inside GNSS. 2009;**2009**:22. Available from: https://www.insidegnss.com/auto/julyaug09-kintner.pdf

[5] Carrano CS, Groves KM, McNeil WJ, Doherty PH. Direct measurement of the residual in the ionosphere-free linear combination during scintillation. In: Proceedings of the 2013 International Technical Meeting of the Institute of Navigation. San Diego, California: Catamaran Resort Hotel; January 29-27, 2013. pp. 585-596

[6] Basu S, Basu S, Khan BK. Model of equatorial scintillation from in situ measurements. Radio Science. 1976;**11**: 821-832

[7] Basu S, Basu S, Hanson WB. The Role of in Situ Measurements in Scintillation Modelling. Washington, DC: Nav. Res. Lab; 1981

[8] Wernik AW, Alfonsi L, Materassi M. Scintillation modeling using in situ data. Radio Science. John Wiley & Sons, Ltd. 2007;**42**. DOI: 10.1029/2006RS003512. Available from: https://agupubs.onlinelibrary.wiley.com/action/showCitFormats?doi=10.1029%2F2006RS003512. ISSN 0048-6604. eISSN 1944-799X

[9] Fremouw EJ, Rino CL. An empirical model for average F: Layer scintillation at VHF/UHF. Radio Science. 1973;**8**: 213-222

[10] Aarons J. Construction of a model of equatorial scintillation intensity. Radio Science. 1985;**20**:397-402

[11] Franke SJ, Liu CH. Modeling of equatorial multifrequency scintillation. Radio Science. 1985;**20**:403-415

[12] Iyer KN, Souza JR, Pathan BM, Abdu MA, Jivani MN, Joshi HP. A model of equatorial and low latitude VHF scintillation in India. Indian Journal of Radio & Space Physics. 2006;**35**:98-104

[13] Retterer JM. Forecasting low-latitude radio scintillation with 3-D ionospheric plume models: 2. Scintillation calculation. Journal of Geophysical Research. 2010;**115**:A03307. DOI: 10.1029/2008JA013840

[14] Liu JY, Chen SP, Yeh WG, Tsai HF, Rajesh PK. The worst-case GPS scintillations on the ground estimated by using radio occultation observations of FORMOSAT-3/COSMIC during 2007–2014. Surveys in Geophysics. 2016;**37**: 791. DOI: 10.1007/s10712-015-9355-x

[15] Chen SP, Bilitza D, Liu J-Y, Caton R, Chang L-C, Yeh W-H. An empirical model of L-band scintillation S4 index constructed by using FORMOSAT-3/COSMIC data. Advances in Space Research. 2017;**60**(5):1015-1028. DOI: 10.1016/j.asr.2017.05.031

[16] Kepkar A, Arras C, Wickert J, Schuh H, Alizadeh M, Tsai L-C. Occurrence climatology of equatorial plasma bubbles derived using Formo Sat-3 / COSMIC GPS radio occultation data. Annales de Geophysique. 2020;**38**: 611-623. DOI: 10.5194/angeo-38-611-2020

[17] Fremouw EJ, Secan JA. Modeling and scientific application of scintillation results. Radio Science. 1984;**19**(3): 687-694. DOI: 10.1029/RS019i003p00687

[18] Secan JA, Bussey RM, Fremouw EJ, Basu S. An improved model of equatorial scintillation. Radio Science. 1995;**30**(3): 607-617. DOI: 10.1029/94RS03172

[19] Secan JA, Bussey RM, Fremouw EJ, Basu S. High-latitude upgrade to the wideband ionospheric scintillation model. Radio Science. 1997; **32**(4):1567-1574. DOI: 10.1029/97RS00453

[20] Rino CL. A power law phase screen model for ionospheric scintillation: 1. Radio Science. 1979;**14**(6):1135-1145. DOI: 10.1029/RS014i006p01135

[21] Rino CL. A power law phase screen model for ionospheric scintillation: 2. Radio Science. 1979;**14**(6):1147-1155. DOI: 10.1029/RS014i006p01147

[22] Cervera MA, Thomas RM, Groves KM, Ramli AG, Effendy. Validation of WBMOD in the Southeast Asian region. Radio Science. 2001;**36**(6):1559-1572. DOI: 10.1029/2000RS002520

[23] Béniguel Y. GIM: A global Ionospheric propagation model for scintillations of transmitted signals. Radio Science. 2002;**37**(3):1032. DOI: 10.1029/2000RS002393

[24] GISM Global Ionospheric Scintillation Model. Technical Manual. Available from: http://www.ieea.fr/help/gism-technical.pdf [Accessed: December 18, 2022]

[25] Vasylyev D, Béniguel Y, Volker W, Kriegel M, Berdermann J. Modeling of ionospheric scintillation. Journal of Space Weather Space Climate. 2022;**12**: 22. DOI: 10.1051/swsc/2022016

[26] Radicella SM. The NeQuick model genesis, uses and evolution. Annals of Geophysics. June/August 2009;**52**(3/4): 417-422. DOI: 10.4401/ag-4597

[27] Lannelongue S et al. Characterization of Scintillation effect on Galileo sensor station Continuity of Service" GNSS08 Toulouse (2008) IEEA publication. 2008. Available online at: www.ieea.fr/publications/ieea-2008-gnss.pdf [Last visited on December 18th, 2022]

[28] Ionospheric scintillations specification for EGNOS V3 (ESA/ESTEC), Ref: TEC-EEP/2013.99/RPC, Issue 1, 27/02/2014

[29] Ionospheric Scintillation Effects in EGNOS Performance Assessment (ESA) Ref: E-TN-PFM-E-0155-ESA, Issue 2, 01/03/2013

[30] Ionospheric effects on EGNOS (ESA/ESTEC) Ref: TEC-EPP/2010.687/RPC, Issue 1, 15/12/2012

[31] Galiègue H, Féral L, Fabbro V. Validity of 2-D electromagnetic approaches to estimate log-amplitude and phase variances due to 3-D ionospheric irregularities. Journal of Geophysical Research: Space Physics. 2017;**122**:1410-1427. DOI: 10.1002/2016JA023233

[32] Shkarofsky IP. Generalized turbulence space-correlation and

wave-number spectrum-function pairs. Canadian Journal of Physics. 1968; **46**(19):2133-2153. DOI: 10.1139/p68-562

[33] Fabbro V, Jacobsen KS, Rougerie S. HAPEE, a prediction model of ionospheric scintillation in polar region. In: 2019 13th European Conference on Antennas and Propagation (EuCAP), Krakow, Poland. 31 March-5 April 2019. pp. 1-5

[34] Camps A et al. Improved modelling of ionospheric disturbances for remote sensing and navigation. IEEE International Geoscience and Remote Sensing Symposium (IGARSS). *2017*; **2017**:2682-2685. DOI: 10.1109/IGARSS.2017.8127548

[35] Bilitza, D. IRI - International Reference Ionosphere. Available from: http://iri.gsfc.nasa.gov/ [Accessed: December 18, 2022]

[36] Rovira-Garcia A et al. Climatology of high and low latitude scintillation in the last solar cycle by means of the geodetic detrending technique. In: Proceedings of the 2020 International Technical Meeting of The Institute of Navigation. San Diego, CA; 2020. pp. 920-933. DOI: 10.33012/2020.17187

[37] Blanch E et al. Improved characterization and modeling of equatorial plasma depletions. Journal of Space Weather Space Climate. 2018;**8**:A38. DOI: 10.1051/swsc/2018026

[38] Juan JM et al. A method for scintillation characterization using geodetic receivers operating at 1 Hz. Journal of Geodesy. 2017;**91**(11): 1383-1397. DOI: 10.1007/s00190-017-1031-0

[39] Nguyen VK et al. Measuring phase scintillation at different frequencies with conventional GNSS receivers operating at 1 Hz. Journal of Geodesy. 2019;**93**: 1985-2001. DOI: 10.1007/s00190-019-01297-z

[40] Sanz J et al. Novel ionospheric activity indicator specifically tailored for GNSS users. In: Proceedings of the 27th International Technical Meeting of The Satellite Division of the Institute of Navigation (ION GNSS+ 2014). Tampa, Florida; 2014

[41] Humphreys TE et al. Modeling the effects of ionospheric scintillation on GPS carrier phase tracking. IEEE Transactions on Aerospace and Electronic Systems. 2010;**46**(4): 1624-1637

[42] Béniguel Y et al. MONITOR Ionospheric Network: Two case studies on scintillation and electron content variability. Annales Geophysicae. 2017; **35**:377-391. DOI: 10.5194/angeo-35-377-2017

[43] Juan JM, Sanz J, Rovira-Garcia A, González-Casado G, Ibáñez-Segura D, Orus R. AATR an ionospheric activity indicator specifically based on GNSS measurements. Journal of Space Weather Space Climate. 2018;**8**(A14): 1-11. DOI: 10.1051/swsc/2017044

[44] Carrano CS, Groves KM, Rino CL. On the relationship between the rate of change of total electron content index (ROTI), irregularity strength (CkL), and the scintillation index (S4). Journal of Geophysical Research: Space Physics. 2019;**124**(3):2099-2112. DOI: 10.1029/2018JA026353

[45] Acharya R, Majumdar S. Statistical relation of scintillation index S4 with ionospheric irregularity index ROTI over Indian equatorial region. Advances in Space Research. 2019;**64**:1019-1033. DOI: 10.1016/j.asr.2019.05.018

Chapter 7

Modeling of the Gradient-Drift Instability in the Earth's Equatorial Ionosphere

Sergey Matsievsky

Abstract

The used mathematical and numerical models of the gradient-drift instability are based on mathematical and numerical models of the Rayleigh-Taylor instability in the Earth's equatorial ionosphere. These models are constructed as systems of electrodynamically matched multidimensional nonlinear equations described equatorial F-layer of the Earth's equatorial ionosphere. These instabilities can generate Earth's plasma irregularities with space-and-time scales that are characteristic of equatorial F-spread. First, the space-and-time model of the developed equatorial plasma bubble is obtained by a two-dimensional numerical simulation of the Rayleigh-Taylor instability. Second, the growth rates of the plasma gradient-drift instability are obtained from the dispersion equation based on physical parameters of the developed equatorial plasma bubble. The results of numerical experiments confirm the possibility of generating the gradient-drift instability of Earth's equatorial ionospheric plasma.

Keywords: gradient-drift instability, Earth's equatorial ionosphere, mathematical modeling, numerical modeling, growth rate, Rayleigh-Taylor instability

1. Introduction

Depleted concentration zones in the Earth's equatorial ionosphere F-layer (named equatorial plasma bubbles, EPBs) are very irregular and nonstationary 3D complex structures. It is of great importance that EPBs have a very scattered range of 3D objects from huge to very little [1–5]. Under favorable conditions in the Earth's equatorial ionospheric plasma for the Rayleigh-Taylor instability (RTI), medium-sized initial irregularities develop in a hyperbolic growth into huge EPBs [6–8]. It should be particularly noted that in this ionospheric process, small irregularities of small sizes do not begin to develop immediately. These small-sized irregularities can develop exclusively on the basis of sufficiently developed EPB and under favorable conditions for a variety of gradient instabilities, for example, the gradient-drift instability.

EPB groups are constantly observed from the outside of the Earth's ionosphere. EPB in these groups differs greatly in size, as well as in the degree of development. Such measurements are also carried out within the framework of the *Defense Meteorological Satellite Program (DMSP)* [9]. Artificial Earth satellites DMSP *provides cloud cover imagery from polar orbits that are Sun-synchronous at a nominal altitude of 830* km (**Figure 1**). The DMSP satellite performs valuable EPB observations, although the satellites on their trajectory are over the Earth's equator for a short time.

In this work, the problem of the real feasibility of the occurrence and further deepening of secondary small-sized plasma irregularities fueled by the *gradient-drift instability (GDI)* [10, 11] at the EPB boundaries is investigated. The existence of such a cascade mechanism is made possible because:

- Huge differences in ionic density at neighboring longitudes occur at the boundaries of deep EPBs.

- Huge differences in ionic density at neighboring heights occur at the boundaries of deep EPBs.

It is difficult to study two instabilities of different scales at once: RTI and GDI. Therefore, the following cascade mechanism was used in this work:

- The plasma density as the basis of the ongoing processes is calculated on a two-dimensional simulation, which is satisfactory due to the condition that the EPB is strongly elongated along the Earth's magnetic field [6, 12].

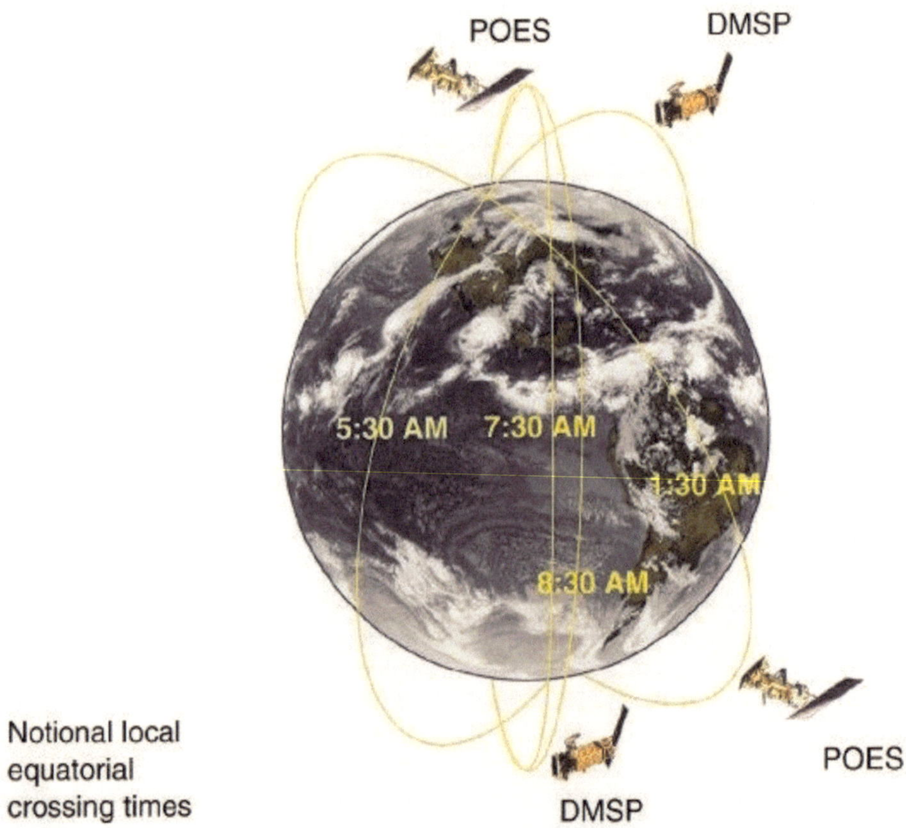

Figure 1.
DPMS orbits. (https://en.wikipedia.org/wiki/File:Operational_polar_satellites.jpg).

DOI: http://dx.doi.org/10.5772/intechopen.1001060

- The real possibility of the emergence and further deepening of GDI is determined by the *growth rate (GR)* of GDI.

2. 3D mathematic model of the Earth's ionosphere

2.1 3D basic eqs

1. The atmosphere of Earth is the layer of gases, and the ionosphere is the ionized part of the upper atmosphere. The gas dynamics can be described in the most detail in the following three ways [12]:

    - The most detailed known description of gas motion consists of the precise determination of the instantaneous positions and velocities of all the molecules that make up the gas.

    - In practice, a molecular distribution function is introduced that satisfies the integrodifferential Boltzmann equation, by which this function is fundamentally determined by its given initial value. However, the description of the distribution function is too detailed.

    - Therefore, in reality, the equations of thermodynamics are used for calculations. In these equations, the parameters of the state are the velocity of the macroscopic mass flow, density, and temperature. In the derivation of these equations, methods dating back to Grad [13] are used.

    Note that in the equilibrium state, there is no difference between these three approximations.

    Plasma, formed at ionospheric heights, is commonly called thermal, thereby emphasizing that the distributions of electrons and ions are the Maxwell-Boltzmann distribution and can be characterized by their temperatures

    The mean free path of thermal plasma particles is usually significantly less than the characteristic spatial scales of changes in its parameters. Therefore, the thermal plasma in the ionosphere is considered as a continuous medium and is used for its modeling by the equations of magnetic hydrodynamics.

    The mathematical base of the three-dimensional numerical ionospheric model of the RTI, denoted by MI3, is the system of equations of plasma gas dynamics containing the following scalar equations for (N: 1) ion types and electrons [14–16]

2. N scalar equations of *ions and electrons continuity* under the assumption of the quasi-neutrality of the ionospheric plasma (MI3/1):

$$\frac{\partial n_j}{\partial t} + \nabla\left(n_j \mathbf{V}_j\right) = Q_j - L_j n_j; \qquad (1)$$

3. N scalar equations of *ions and electrons motion* (MI3/2):

$$\frac{\partial \mathbf{V}_j}{\partial t} + \mathbf{V}_j \nabla \mathbf{V}_j = \frac{e_j}{m_j}\left(\mathbf{E} + \mathbf{V}_j \times \mathbf{B}\right) - \frac{\nabla p_j}{n_j m_j} + \mathbf{g} - \nu_{jn}\left(\mathbf{V}_j - \mathbf{V}_n\right) - \sum_{j \neq k} \nu_{jk}\left(\mathbf{V}_j - \mathbf{V}_k\right); \quad (2)$$

4. N scalar equations of *ions and electrons thermal conductivity* (MI3/3):

$$\frac{3}{2}n_j k\left(\frac{dT_j}{dt}+(\mathbf{V}_j\nabla)T_j\right)+p_j\nabla\mathbf{V}_j+\nabla q_j=G_j-P_j; \tag{3}$$

5. Three scalar equations of *electric field potentiality* (MI3/4):

$$\nabla\times\mathbf{E}=0; \tag{4}$$

6. Three scalar equations of *electric current continuity* (MI3/5):

$$\nabla J\equiv\nabla\sum e_j n_j\mathbf{V}_j=0, \tag{5}$$

where j is *ion and electron type*, (N – 1) ions; Q_j and L_j are *ion and electron addition* and *loss rates*, respectively; $\mathbf{V}_j$, n_j, m_j, e_j, and p_j are *drift velocity*, *density*, *mass*, *charge*, and *gas pressure* of *ions and electrons*, respectively; ν_{jn} and ν_{jk} are *ions and electrons collision frequency with neutrals* and *with each other*, respectively; T_j are *ion and electron temperature*; G_j and P_j are *heating and cooling rates*, respectively; q_j is *heat flux density*; the subscripts indicate the *ion and electron* types; k is *Boltzmann's constant*; $\mathbf{E}$ is *electric field strength*; and J is *electric current density*.

2.2 Dipole coordinate system

1. In the study area, which is in the altitudinal range from 150 to 950 km, it is possible to make the following assumptions about the transport processes due to the strong magnetization of the ionospheric plasma:

   - Transport processes are determined by collisions along the magnetic field.
   - Transport processes are determined by the drift motion of the plasma across the field.

   Due to the strong anisotropy caused by the Earth's magnetic field, the processes of diffusion transfer and thermal conductivity in region F of the equatorial ionosphere occur mainly along the lines of force of the geomagnetic field. With a successful choice of the coordinate system, the cumbersome equations and boundary conditions in general will be simplified.

2. Let us choose a dipole coordinate system for the reason that the Earth's magnetic field is approximated by the dipole approximation [17]. The coordinate components (α, φ, and β) of a curved orthogonal dipolar system are defined in terms of the spherical coordinates of the central dipole as follows:

$$\alpha=\frac{r}{\sin^2\theta},\beta=\frac{\cos\theta}{r^2},\varphi=\varphi, \tag{6}$$

   where r is the distance to the center of the Earth, θ is the latitude, and φ is the longitude (**Figure 2**).

3. Now let $\mathbf{e}_\alpha$, $\mathbf{e}_\beta$, and $\mathbf{e}_\varphi$ be the coordinate orts in the dipolar system, and ort $\mathbf{e}_\beta$ coincides with the direction of the geomagnetic field, ort $\mathbf{e}_\varphi$ is directed from west

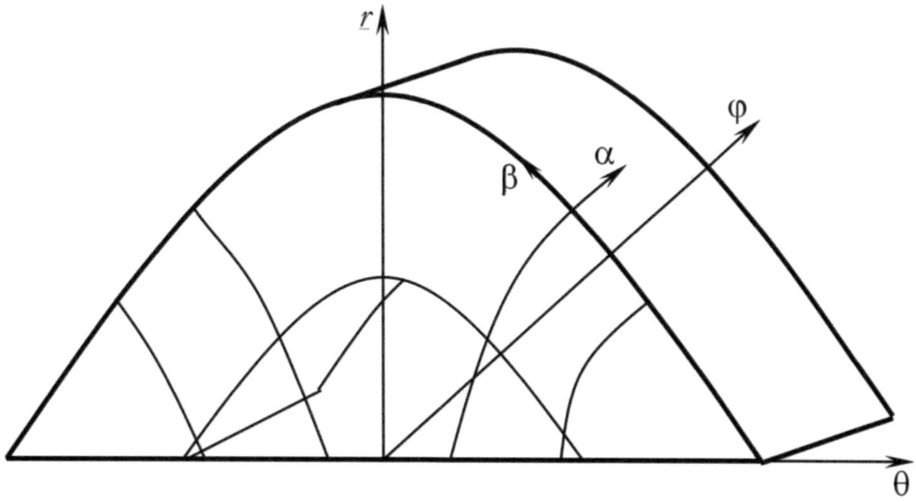

Figure 2.
Dipole coordinate system.

to east, and ort $\mathbf{e}_\alpha$ is chosen so that it is directed upwards at the magnetic equator. Then, we have the left-handed coordinate system:

$$\mathbf{e}_\alpha \times \mathbf{e}_\varphi = \mathbf{e}_\beta, \mathbf{e}_\beta \times \mathbf{e}_\alpha = \mathbf{e}_\varphi, \mathbf{e}_\varphi \times \mathbf{e}_\beta = \mathbf{e}_\alpha. \tag{7}$$

Lame coefficients:

$$H_\alpha = \frac{\sin^3\theta}{\sqrt{\delta}}, H_\varphi = r\sin\theta, H_\beta = \frac{r^3}{\sqrt{\delta}}, \tag{8}$$

where $\delta = 1 + 3\cos^2\theta = 1 + 3\beta^2 r^4$ and $H_\beta = \alpha^2 H_\alpha H_\varphi$. We also denote the product of the Lame coefficients by $H = H_\alpha H_\beta H_\varphi$.

2.3 Reducing the number of differential eqs

1. Simplify the system of equations by reducing the number of differential equations and the number of unknown functions, respectively [16]. For this purpose, the following conditions of the diffusion approximation are used:

$$\frac{d\mathbf{V}_j}{dt} \equiv \frac{\partial \mathbf{V}_j}{\partial t} + \mathbf{V}_j \nabla \mathbf{V}_j = 0. \tag{9}$$

Then

$$\mathbf{V}_j = \hat{\upsilon}_j \left(\mathbf{E} + \frac{B}{a_j} \mathbf{A}_j \right), \tag{10}$$

where $\hat{\upsilon}_j = \begin{pmatrix} \upsilon_{jp} & -\upsilon_{jh} & 0 \\ \upsilon_{jh} & \upsilon_{jp} & 0 \\ 0 & 0 & \upsilon_{j0} \end{pmatrix}$ is the tensor of integral mobility; $\upsilon_{j0} = \frac{a_j'}{B}$;

$\upsilon_{jp} = \frac{a_j'}{B\left(1+a_j'^2\right)}$; and $\upsilon_{jh} = -\frac{a_j'^2}{B\left(1+a_j'^2\right)}$ are longitudinal, Pedersen, and Hall

components, respectively; $B = |\mathbf{B}|$; $a_j = e_j \frac{B}{m_j \nu_{jn}}$; $a'_j = a_j \frac{\nu_{jn}}{\nu_{jn}}$; $\nu_{jn} = \nu_{jn} + \nu_t$;
$\mathbf{A}_j = -\frac{\nabla p_j}{n_j m_j \nu_{jn}} + \frac{\mathbf{g}}{\nu_{jn}} + \mathbf{V}_n$.

Similarly, it obtains that

$$\mathbf{j} = \hat{\sigma}\mathbf{E} + \mathbf{A}, \tag{11}$$

where $\hat{\sigma} = \sum e_j n_j \hat{\upsilon}_j \equiv \begin{pmatrix} \sigma_p & -\sigma_h & 0 \\ \sigma_h & \sigma_p & 0 \\ 0 & 0 & \sigma_0 \end{pmatrix}$ is the tensor of integral conductivity across the magnetic field; $\mathbf{A} = B \sum e_j n_j \hat{\upsilon}_j \frac{\mathbf{A}_j}{a_j}$.

2. The following simplification is related to the rearrangement of the electron density in the ionosphere. Estimates [18] show that the neutrality of ionospheric plasma is maintained with high accuracy. This is enough to accept the condition of neutrality in the equations of motion and continuity of densities. Therefore

$$\sum e_j n_j = 0, \tag{12}$$

moreover, the process of establishing a quasi-neutral state does not need to be investigated.

In addition, the magnetic field is assumed to be constant in time

$$\mathbf{B} = \text{const} \tag{13}$$

and dipole. This is possible because the variations of the geomagnetic field created by electric currents in the irregularity and in the ionosphere as a whole are negligible compared to the geomagnetic field [18].

3. Due to the condition of electrostatics

$$\nabla \times \mathbf{E} = 0, \tag{14}$$

the electric field is potentially (MI3/6):

$$\mathbf{E} = -\nabla \mathbf{\Phi}, \tag{15}$$

$\mathbf{\Phi}$ is the electrical potential.

The dipole model geomagnetic field will not change during calculations, which makes it possible to go from 3D MI3/5 to 2D MI3/7 by integrating the *electric current continuity equation*:

$$\nabla_\perp(\hat{\sigma}\nabla_\perp\Phi) = \nabla_\perp\mathbf{A}, \tag{16}$$

$\nabla_\perp$ acts perpendicular to the force lines.

Since the electric current density, as well as the velocities of charged particles, can be calculated based on algebraic expressions, they are also not included in the final system.

The result is $2N$ three-dimensional nonstationary nonlinear equations with $2N$ unknowns $n_2, \ldots, n_N, T_1, \ldots, T_N, \Phi$:

$$\begin{cases} \frac{\partial n_i}{\partial t} + \nabla(n_i \mathbf{V}_i) = Q_i - L_i, \ i = \overline{2,\ N}, \\ \frac{3}{2} n_j k \frac{dT_j}{dt} + p_j \nabla \mathbf{V}_j + \nabla q_j = G_j - P_j, \\ j = \overline{1,\ N}, \\ \nabla_\perp (\hat{\sigma} \nabla_\perp \Phi) = \nabla_\perp \mathbf{A}. \end{cases} \quad (17)$$

2.4 3D final eqs

1. The resulting system of nonlinear three-dimensional strongly coupled multicomponent hydrodynamic equations with nonlocal properties allows for further simplification. The greatest difficulties to solve are the three-dimensional inhomogeneous asymmetric elliptic equation for the electric field potential obtained from the electric current continuity equation. Therefore, this equation will be simplified.

   First, let us take advantage of the fact that there is a dedicated direction in the ionospheric plasma—the direction of the geomagnetic field. The plasma along it at the heights of the equatorial region F is a highly conductive medium due to a large amount of longitudinal conductivity σ_0. Therefore, let the geomagnetic field lines be equipotential, then the electric potential will be independent of the direction of the geomagnetic field

$$\Phi = \Phi(\alpha,\ \varphi). \quad (18)$$

2. Finally, the dimension of the three-dimensional equation for the potential is reduced by integrating it along the lines of force. At the same time, the integration region should take into account the parameters of the E-region: its lines of force should begin under the E-region in the southern hemisphere, pass through the geomagnetic equator, and end under the E-region in the northern hemisphere (see **Figure 2**).

   Then, at the ends of the power lines, natural boundary conditions can be set, consisting of the impermeability of the lower boundary of the E-region to the electric current. As a result, the integral of the derivative in the direction of the geomagnetic field turns to zero, and only the following integral remains:

$$\int_{\mathrm{E+F}} H \nabla_\perp \mathbf{j} d\beta = 0, \quad (19)$$

   where E and F are E- and F-regions.

3. So, a three-dimensional model MI3 suitable for numerical solution is constructed

$$\begin{cases} \dfrac{\partial n_i}{\partial t} + \nabla(n_i \mathbf{V}_i) = Q_i - L_i, \ \ i = \overline{1, N}, \\ \dfrac{3}{2} n_j k \dfrac{dT_j}{dt} + p_j \nabla \mathbf{V}_j + \nabla q_j = G_j - P_j, \ \ j = e, \ i, \\ \dfrac{\partial \left(\int\limits_{\mathrm{E+F}} \sigma_P \dfrac{H}{H_\alpha^2} d\beta \cdot \dfrac{\partial \Phi}{\partial \alpha} \right)}{\partial \alpha} + \dfrac{\partial \left(\int\limits_{\mathrm{E+F}} \sigma_P \dfrac{H}{H_\varphi^2} d\beta \cdot \dfrac{\partial \Phi}{\partial \varphi} \right)}{\partial \varphi} + \\ \dfrac{\partial \int\limits_{\mathrm{E+F}} \sigma_h H_\beta d\beta}{\partial \alpha} \cdot \dfrac{\partial \Phi}{\partial \varphi} - \dfrac{\partial \int\limits_{\mathrm{E+F}} \sigma_h H_\beta d\beta}{\partial \varphi} \cdot \dfrac{\partial \Phi}{\partial \alpha} = \int\limits_{\mathrm{E+F}} \nabla_\perp \mathbf{A} d\beta. \end{cases} \quad (20)$$

The well-known *global empirical model MSIS* was required to determine the characteristics of the uncharged component of the Earth's ionosphere [19, 20].

Before starting calculations of the development of irregularities, the model was established by running a background simulation of ionospheric plasma until it reached a periodic solution with a period of a day.

This 3D MI3 model has been successfully used (e.g., [21]).

3. 2D mathematic models of the Earth's ionosphere

1. The following simplification is related to the following property of the majority of sufficiently developed medium-scale equatorial irregularities obtained from observational data and numerical three-dimensional experiments on the previous three-dimensional model. Irregularities are strongly elongated along the lines of force and remain so during their movement. Let the irregularities be two-dimensional, then their dynamics are described in the plane of the geomagnetic equator by two-dimensional equations.

   By discarding the terms with longitudinal components in the equations of the three-dimensional model and translating all variables to the coordinates α and φ, a **two-dimensional model MI2** of developed irregularities is obtained, suitable for numerical solution:

$$\begin{cases} \dfrac{\partial n_i}{\partial t} + \nabla_\perp(n_i \mathbf{V}_i) = Q_i - L_i, \ \ i = \overline{1, N}, \\ \dfrac{3}{2} n_j k \dfrac{dT_j}{dt} + p_j \nabla_\perp \mathbf{V}_j + \nabla_\perp q_j = G_j - P_j, \\ j = e, \ i, \\ \nabla_\perp(\hat{\sigma} \nabla_\perp \Phi) = \nabla_\perp \mathbf{A}. \end{cases} \quad (21)$$

   This 2D MI2 model has been successfully used (e.g., [6]).

2. The last significant simplification is based on observational data and numerical experiments on previous models, which show that sufficiently developed

irregularities are strongly elongated in height. But this circumstance is used only in the equation for the potential, where only the zonal gradient is taken into account:

$$\frac{\partial j_\varphi}{\partial \varphi} = 0. \tag{22}$$

In addition, let $N = 1$, considering only one positive ion, which usually does not lead to incorrect conclusions in the dynamics of irregularities [14].3. Because due to the magnetization of the plasma of the F-layer

$$a_1 = e_1 \frac{B}{m_1 \nu_{1n}}, \tag{23}$$

then

$$\frac{\partial j_\varphi}{\partial \varphi} = \frac{\partial(\sigma_p E_\varphi + A_\varphi)}{\partial \varphi} = 0, \sigma_p E_\varphi + A_\varphi = \text{const } (\alpha) = j_{0\varphi}(\alpha), \tag{24}$$

and

$$E_\varphi = \frac{j_{0\varphi}(\alpha) - A_\varphi}{\sigma_p}, \tag{25}$$

where $j_{0\varphi}(\alpha)$ is the longitude component of the background electric current $\mathbf{j}_0 = \mathbf{A}$, which must depend on the height only.4. In the simplest case

$$j_{0\varphi} = A_\varphi \cong n_{10} m_1 \frac{g}{B}, \tag{26}$$

where $n_{10}(\alpha)$ is the background altitude profile of the ion concentration. Now the zonal component of the electric field E_φ can be found from the algebraic expression.

The following four equations with four unknowns n_1, T_1, T_e, E_α form an **express model MI0** of developed irregularities, suitable for both numerical and analytical solutions:

$$\begin{cases} \dfrac{\partial n_1}{\partial t} + \nabla_\perp(n_1 \mathbf{V}_1) = Q_1 - L_1, \\ \dfrac{3}{2} n_j k \dfrac{dT_j}{dt} + p_j \nabla_\perp \mathbf{V}_j + \nabla_\perp q_j = G_j - P_j, \ j = e, \ i, \\ \dfrac{\partial E_\alpha}{\partial \varphi} = \dfrac{\partial E_\varphi}{\partial \alpha}. \end{cases} \tag{27}$$

This 2D MI0 model has been successfully used (e.g., [22]).

An analytical solution of the 2D MI0 model was obtained [23].

4. Gradient-drift instability model

4.1 Mathematic model

1. The two-dimensional model MI2 is used.

   When a deep EPB is formed, very large density differences of charged plasma particles occur on the entire contour of this EPB over the Earth's magnetic field:

   - Longitude very large density differences of charged plasma particles.

   - Altitude very large density differences of charged plasma particles.

   These huge differences in plasma density favor the emergence and intensification of *drift waves* growing in the hyperbolic growth regime, when the linear increment is sufficiently large. **Figure 3** shows schematic drawing of the strongly deep EPB structure in the plane of the Earth's equator with numbers:

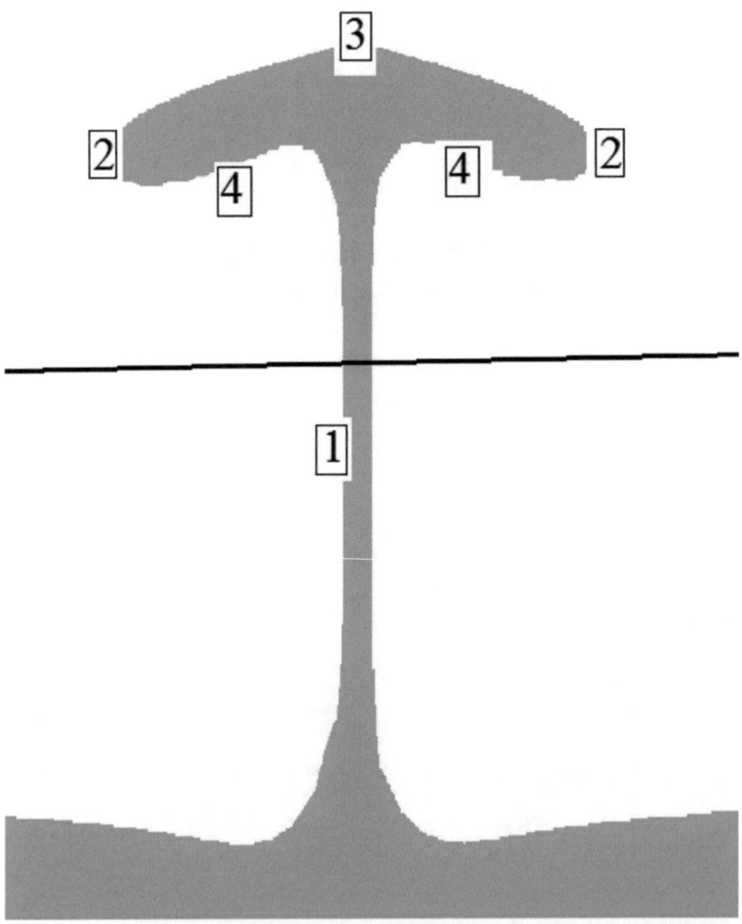

Figure 3.
Schematic structure of the developed EPB.

- he horizontal straight line across EPB indicates the maximum of the background F-layer of the Earth's ionosphere.
- "1" is located at the place of the greatest drop in plasma density.
- "2," "3," and "4" are in place of the lateral, upper, and lower boundaries with huge drops in plasma density, respectively.

2. To be able to work with the GDI, two equations are added. First, the following notations are introduced: N_0 is the unperturbed electron concentration; Φ_0 is the unperturbed potential of the electric field, made dimensionless by dividing by the induction of the magnetic field B; $V_{0\varphi}$ and $V_{0\alpha}$ are the undisturbed drift speeds; and ν_r is the recombination rate of electrons. The perturbed values are also denoted but without index 0 [24].

   Then for the perturbed values N and Φ taking into account the fact that the conditions $\nabla_\perp(V_\varphi) = 0$ and $\nabla_\perp(V_\alpha) = 0$ for the ionospheric plasma are satisfied, two equations are added:

$$\frac{\partial N}{\partial t} + V_{0\varphi}\frac{\partial N}{\partial \varphi} + V_{0\alpha}\frac{\partial N}{\partial \alpha} + \nu_r N + \frac{\partial \Phi}{\partial \varphi}\frac{\partial N_0}{\partial \alpha} - \frac{\partial \Phi}{\partial \alpha}\frac{\partial N_0}{\partial \varphi} + \frac{\partial \Phi}{\partial \varphi}\frac{\partial N}{\partial \alpha} - \frac{\partial \Phi}{\partial \alpha}\frac{\partial N}{\partial \varphi} = 0,$$
$$\Delta\Phi + \frac{V_{0\varphi}}{N_0}\frac{\partial N}{\partial \alpha} + \frac{V_{0\alpha}}{N_0}\frac{\partial N}{\partial \varphi} = 0. \quad (28)$$

   The system of two differential equations is nonlinear since the first equation is nonlinear. The system of these two differential equations can be solved using only the first two harmonics of variation:

$$N = N_1 \exp(-i\omega t + ik_\varphi \varphi + ik_\alpha \alpha) + N_2 \exp(-2i\omega t + ik_\varphi \varphi + ik_\alpha \alpha) +$$
$$N_3 \exp(-i\omega t + 2ik_\varphi \varphi + 2ik_\alpha \alpha),$$
$$\Phi = \Phi_1 \exp(-i\omega t + ik_\varphi \varphi + ik_\alpha \alpha) + \Phi_2 \exp(-2i\omega t + ik_\varphi \varphi + ik_\alpha \alpha) +$$
$$\Phi_3 \exp(-i\omega t + 2ik_\varphi \varphi + 2ik_\alpha \alpha). \quad (29)$$

3. In accordance with refs. [10, 11], under the condition $k_\varphi < < k_\alpha$, the *instability increment* obtained from this system of two equations has the form

$$\gamma_g = \frac{V_{0\varphi}}{L_\varphi} \cdot \frac{k_\alpha^2}{k_\varphi^2 + k_\alpha^2} - \nu_r, \; L_\varphi = \left(\frac{\partial \ln N_0}{\partial \varphi}\right)^{-1}. \quad (30)$$

Such a nonlinear increment of increase can be used in the case when an irregularity of the plasma density occurs at the lateral boundaries of the deep EPB. The frequencies of plasma waves in the presented mathematical simulation should not be less than the frequencies corresponding to ionospheric irregularities at the boundaries of which they exist. Also, these frequencies do not rise above the wave frequencies of the charged components of the ionosphere. In this case, we have wave sizes of 10–1000 m.

4.2 Numerical model

1. The computational simulation occupied the plane of the magnetic equator not lower than 100 km and not higher than 1700 km. In width, the two-dimensional simulation space occupied approximately 400 km. A relatively large two-dimensional simulation space serves to exclude the impact of processes at the edges on the results of simulation of the development of irregularities. Numerical simulation was carried out at medium $F_{10.7} = 150$ and small $k_p = 3$ [24].

   The boundary values of the electric potential are obtained from the undisturbed electric potential:

   - On the right, the model's undisturbed electric potential was only 0.001 V/m. This value sets the movement from the bottom up with a model speed of 40 m/s.

   - At the bottom of the simulation area, the density of charged particles corresponds to the chemical equilibrium of the processes, and the heating of charged particles is equivalent to the heating of neutral plasma particles.

   - The model value of 0 is taken from above and on the sides for the plasma particle flow rates.

   The model seed EPB is given by the following equation:

$$n_i = n_{i0}\left(1 + (a_n - 1)\exp\left(-\left(\frac{\varphi - \varphi_0}{l}\right)^2 - \left(\frac{\alpha - \alpha_0}{l}\right)^2\right)\right), \qquad (31)$$

   where n_{i0} is the background concentration of ions; a_n is the ratio of the concentration at the center of the irregularity to the background value of the concentration at this height ($a_n = 0.5$ for the simulation); $(\varphi_0, \ \alpha_0)$ are the model seed EPB center coordinates; and l is the model seed EPB radius ($l = 5$ km for the simulation).

2. The system of differential equations of the mathematical model was approximately solved by *approximating derivatives with finite difference* on *quasi-uniform grids*, grinding to the middle of the two-dimensional simulation space to the size of 1.5 km. The solution of the *equations of motion* has been simplified by the fact that they are *algebraic*. The solution of the equation for the electric potential was difficult, because it is not just elliptical, but also:

   - It is not self-adjoint due to the presence of *Hall conductivity* components in the equation.

   - It significantly depends on the ion density of the ionosphere.

   Due to the complexity of solving finite-difference equations, it was necessary to develop approximate second-order equations of accuracy. On the other hand, the use of only two-dimensional simulation of processes in the ionosphere allowed the use of small finite-difference grids, which made it possible to satisfactorily calculate the little scale of some parts of the EPB.

Numerical solution of mathematical models of Rayleigh-Taylor instability has some difficulties that not all authors manage to overcome [25]:

- Computational *finite-difference schemes* for solving *transfer equations* in numerical experiments with RTI simulation should have satisfactory accuracy (second order).

- The traditional erroneous way, practiced by some authors for more than half a century, is to neglect the fact that RTI develops in a mode with aggravation. Such a development of events with a poor choice of *finite-difference schemes* is the reason for a critical increase in the errors of approximate calculations, which is the reason for the nonphysical results of numerical experiments.

3. The finite-difference *transfer equations* included in the numerical model are solved as follows

   - Nonlinear 2D equations are solved according to the so-called *splitting scheme*, and this scheme is symmetric, as a result of which the scheme has the *second order of accuracy*.

   - The 1D equation is solved by a specially developed *monotonic method with nonlinear correction of flows* for which the *minmod limiter* was chosen [26, 27].

   The *electric potential equation* included in the numerical model is solved using the *multigrid method* with the specially developed *W-scheme*.

   According to numerous calculations by various authors, the deep EPB has the following characteristics:

   - The width of the "leg" of the mushroom-shaped EPB is approximately $L_\varphi \approx 1\ \text{km}$.

   - The width of the "cap" of the mushroom-shaped EPB is approximately $L_\varphi \approx 10 - 100\ \text{km}$.

A numerical experiment to simulate the development of RTI made possible that both the charged particles *drift velocity* and the *background irregularities* sizes are obtained.

4.3 Results of numerical experiments

1. The finite-difference simulation MI2 of the RTI development constructed above was the first step necessary to determine the values of the simulated deep EPB parameters of the Earth ionosphere and their differences, which, in turn, are necessary to calculate the values of the GDI development rate. **Figure 4** presents the calculated contours of the ionospheric components with the following characteristics:

   - The numerical experiment was carried out on a rectangle with a width of 100 km (longitude) and a height of 800 km (from 150 to 950 km).

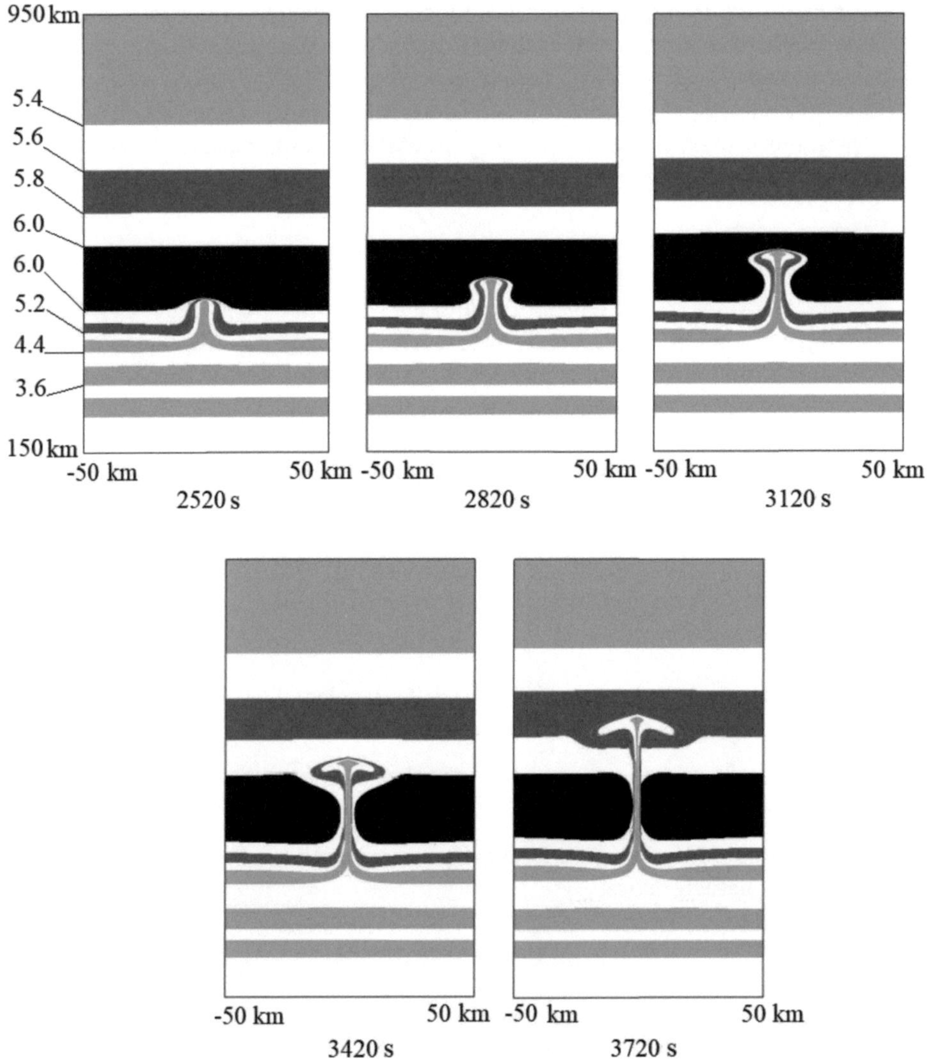

Figure 4.
Distribution of electron concentration at different times from the beginning of initialization in the plane of the magnetic equator. The marks on the left at the boundaries of the zones are equal to the decimal logarithm of the electron concentration.

- The process snapshots are shown for the times 2500, 2800, 3100, 3400, and 3700 s, respectively, since the launch of RTI by creating the model seed EPB [24].

2. **Figure 5** represents the values of the following parameters, which were obtained during a computational experiment using mathematical-numerical simulation of MI2, varying in time, which is counted since the launch of RTI by creating the model seed EPB

- **Figure 5(a)** represents the calculated maxima of the GDI *growth increment* $1000\gamma_g$ (1/s).

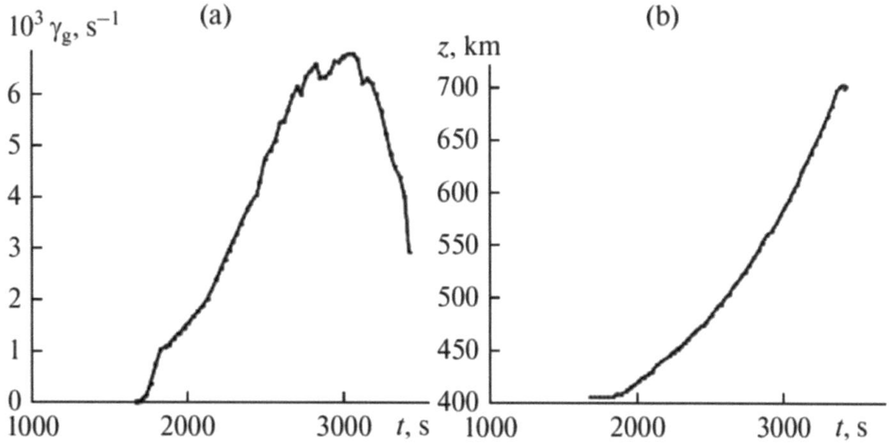

Figure 5.
The values calculated by model MI2 and depending on time from the moment of initialization of RTI: (a) maximum value of the GDI growth increment $1000\gamma_g$ *1/s and (b) the height of the maximum value* γ_g *km.*

- **Figure 5(b)** represents the calculated height of the maximal value of the GDI *growth increment* γ_g (km).

3. In accordance with **Figure 5**, which shows the results of a numerical experiment, the calculated linear growth rate of GDI can rise to values of 0.006 1/s, or the characteristic linear development time of GDI can be 170 s. It is obvious that the obtained values of the linear increment of GDI may well be the reason for the rather rapid growth of irregularities of small size, which, in fact, are the main factor in the development of F-spread in the equatorial latitudes of the Earth at ionospheric heights.

5. Conclusions

Using the MI2 model for numerical calculations on a personal computer leads to the following conclusions about the nature of F-spread:

1. Density differences of charged particles of equatorial ionospheric plasma across the Earth's magnetic field at the lateral boundaries of the cap of a deep mushroom-like EPB can most likely be the initial link of the cascade mechanism from density differences to GDI.

2. The linear growth rate of secondary GDI may well rise to values of 1 /(170 s) and higher, being the trigger for the appearance and deepening of irregularities of small size.

3. The constructed mathematical and numerical simulation is designed to simulate irregularities of tens of meters in size, so the numerical experiment showed that GDI could well provoke equatorial F-spread.

Author details

Sergey Matsievsky
Independent Researcher, Kaliningrad, Russia

*Address all correspondence to: sergei.matsievsky@yandex.ru

References

[1] Farley DT, Balsley BB, Woodman RF, McClure JP. Equatorial spread F: Implications of VHF radar observations. Journal of Geophysical Research. 1970; **75**(34):7199-7216

[2] Reinisch BW, Abdu M, Batista I, et al. Multistation digisonde observations of equatorial spread F in South America. Annales de Geophysique. 2004;**22**: 3145-3153

[3] Rottger J. The macro-scale structure of equatorial spread-F irregularities. Journal of Atmospheric and Terrestrial Physics. 1976;**38**(1):97-101

[4] Saito S, Maruyama T, Ishii M, et al. Observations of small- to large-scale ionospheric irregularities associated with plasma bubbles with a transequatorial HF propagation experiment and spaced GPS receivers. Journal of Geophysical Research. 2008;**113**(A12):1-10

[5] Woodman RF, Spread F. An old equatorial aeronomy problem finally resolved? Annales de Geophysique. 2009;**27**:1915-1934

[6] Matsievskii SV, Kashchenko NM, Nikitin MA. Ionospheric bubbles: Ion composition, velocities of plasma motion and structure. Radiophysics and Quantum Electronics. 1989;**32**(11): 969-974

[7] Huba JD, Joyce G, Krall J. Three-dimensional modeling of equatorial spread F. In: Abdu M, Pancheva D, editors. Aeronomy of the Earth's Atmosphere and Ionosphere, IAGA Special Sopron Book Series. Vol. 2. Dordrecht: Springer; 2011. pp. 211-218

[8] Ossakow SL, Chaturvedi PK. Morphological studies of rising equatorial spread F bubbles. Journal of Geophysical Research. 1978;**83**(A5): 2085-2090

[9] Hanson WB, Coley WR, Heelis RA, Urquhart AL. Fast equatorial bubbles. Journal of Geophysical Research. 1997; **102**(A2):2039-2045

[10] Sukovatov YA. Numerical modeling of two-dimensional nonlinear gradient-drift instability in the outer ionosphere. Izv. Altai. Gos. Univ. 2011;**1-2**:168-171

[11] Sukovatov YA. The theoretical study on nonlinear gradient-drift instability in the outer ionosphere. Izv. Altai. Gos. Univ. 2012;**1-1**:222-225

[12] Kashchenko NM, Ishanov SA, Matsievsky SV. Rayleigh-Taylor instability development in the equatorial ionosphere and a geometry of an initial irregularity. Mathematical Models and Computer Simulations. 2019;**11**(3): 341-348

[13] Grad H. On the kinetic theory of rarefied gases. Communications on Pure and Applied Mathematics. 1949;**2**: 331-407

[14] Gershman BN. Dinamika Ionosfernoi Plazmy [Dynamics of Ionospheric Plasma]. Moscow: Nauka; 1974. p. 256. (in Russian)

[15] Kashchenko NM, Matsievsky S. Matematicheskoe modelirovanie neustoichivostei ekvatorialnogo F-sloia ionosfery [Mathematical modeling of instabilities of equatorial F-layer of ionosphere]. Vestnik Kaliningradskogo gosudarstvennogo universiteta. 2003;**3**: 59-68 (in Russian)

[16] Matsievsky SV, Kashchenko NM, Ishanov SA, Zinin LV. 3D-modelirovanie ekvatorialnogo F-rasseianiia: Sravnenie

modelei MI3 i SAMI3 [3D-simulation of the equatorial F-spread: A comparing MI3 with SAMI3]. Vestnik Baltiiskogo federal'nogo universiteta im. I. Kanta. 2013;**4**:102-105 (in Russian)

[17] Fatkullin MN, Sitnov YS. Dipoliarnaia sistema koordinat i ee nekotorye osobennosti [Dipolar coordinates and its some features]. Geomagnetizm i aeronomiia. 1972;**12**(2): 333-335 (in Russian)

[18] Rybin VV, Polyakov VM. Ob ambipoliarnosti dvizhenii ionosfernoi pkasmy [on ambipolarity of ionospheric plasma motions]. Ionosf. Issled. 1983;**33**: 5-44

[19] Hedin AE, Salah JE, Evans JV, et al. A global thermospheric model based on mass spectrometer and incoherent scatter data MSIS. 1 N2 density and temperature. Journal of Geophysical Research. 1977;**82**(16):2139-2147

[20] Hedin AE, Reber CA, Newton GP, et al. A global thermospheric model based on mass spectrometer and incoherent scatter data MSIS. 2 composition. Journal of Geophysical Research. 1977;**82**(16):2148-2156

[21] Gaidukov VY, Kashchenko NM, Matsievskii SV, et al. Zapusk ekvatorialnykh puzyrei putiem modifikatsii E-sloia [launching equatorial bubbles by E-layer modification]. Geomagnetism and Aeronomy. 1991;**31**(6):1042-1048 (in Russian)

[22] Matsievsky SV, Nikitin MA, Pets AV. O nelineinoi stadii razvitiia neustoichivosti Raleia-Teilora v ekvatorialnoi F-oblasti [On the nonlinear stage of development of Rayleigh-Taylor instability in the equatorial F-region]. Geomagnetism and Aeronomy. 1991; **27**(6):921-924 (in Russian)

[23] Komarov VN, Sazonov SV. Ob analiticheskom podkhode k issledovaniiu Raleia-Teilorovskikh struktur ekvatorialnoi F-oblasti [On an analytical approach to the study of Rayleigh-Taylor structures of the equatorial F-region]. Geomagnetism and Aeronomy. 1991; **31**(6):1032-1036 (in Russian)

[24] Kashchenko NM, Ishanov SA, Matsievsky SV. Numerical study of the gradient-drift instability's growth rate at the fronts of equatorial plasma bubbles. Mathematical Models and Computer Simulations. 2021;**13**(4):623-630

[25] Kashchenko NM, Ishanov SA, Matsievsky SV. Efficient algorithms of the numerical simulation of medium-scale irregularities in a low-latitude ionosphere. Mathematical Models and Computer Simulations. 2017;**9**(6): 743-749

[26] Ladonkina ME, Neklyudova OA, Tishkin VF, Chevanin VS. A version of essentially nonoscillatory high-order accurate difference schemes for systems of conservation laws. Mathematical Models and Computer Simulations. 2010;**2**(3):304-316

[27] Safronov AV. Accuracy estimation and comparative analysis of difference schemes of high-order approximation. Vychisl. Metody Program. (Numer. Methods Program.). 2010;**11**(1):137-143

Chapter 8

Narrowband Stimulated Radiation during Ionospheric Heating Experiments: Recent Observations, Theory, and Modeling

Wayne Scales and Augustine Yellu

Abstract

Ground-based transmission of high-power high frequency (HF) radio waves into the Earth's ionosphere produces a plethora of physical phenomena that result from complex nonlinear processes. This field has commonly been referred to as ionospheric heating or ionospheric modification in pertinent literature, and it has a long, rich history dating back at least to the 1960s. The phenomenon induced by the interaction of a radio wave and ionospheric plasma may serve as a laboratory for investigating fundamental phenomena as well as facilitate ionospheric diagnostics. This chapter will concentrate on a specific phenomenon which is stimulated secondary radiation that exists in the frequency band typically within about 1 KHz of the transmit frequency. This narrowband stimulated radiation has been studied extensively over the past decade or so at the High Frequency Active Auroral Research Program (HAARP) Facility in Gakona, Alaska. HAARP's unprecedented capabilities in terms of maximum power, transmit beam and frequency agility, have resulted in discovery of several new nonlinear ionospheric phenomena that may be leveraged for new diagnostics by using the narrowband stimulated radiation spectrum. This chapter will provide a synopsis of the most recent observations, theory, and advanced computational modeling and provide a future outlook on possibilities for future avenues of investigation.

Keywords: stimulated electromagnetic emissions, second harmonic generation, parametric instabilities, stimulated Brillouin scatter, electrostatic ion cyclotron wave, ion-Bernstein waves, ionospheric diagnostics, ionospheric modification, ionospheric heating

1. Introduction

When a high-power high frequency (HF) radio wave, also typically referred to a pump wave, is injected into, and interacts with the Earth's ionosphere, experiments commonly referred to as ionospheric heating experiments, a plethora of physical processes occur [1]. One of these processes is secondary radiation, produced in the ionosphere, called Stimulated Electromagnetic Emissions (SEEs) which are observed by ground-based receivers typically within 100 km of the ground-based high power ionospheric heating facility. The possibility of generating SEEs during ionospheric heating experiments was theoretically proposed in the late 1970's and SEEs were first observed using the heater at the facility now called the European Incoherent Scatter Radar (EISCAT) heating facility as reported in the early 1980's [2]. Since then, SEEs have been developed into a useful diagnostic for ionospheric parameters as well as the nonlinear processes produced during heating by using the spectral line characteristics and their temporal evolution [3]. SEEs along with other diagnostics probing the heated region such as optics, coherent and incoherent radar scatter, GPS scintillation measurements, and in situ satellite measurements can form an even more effective diagnostic suite for analyzing the ionosphere during these experiments. The spectrum of physical processes occurring during ionospheric heating include electron acceleration, airglow, production of a relatively broad spatial scale of ionospheric irregularities as well as secondary radiation, that is, SEEs. SEEs spectral lines studied in the past occur within a frequency bandwidth of approximately ± 100 kHz around the pump wave frequency ω_0 which typically ranges from 2 to 8 MHz. An effective radiated power (ERP) of 100's of MW to 1 GW is usually available from the pump wave. The ionospheric interaction typically occurs in the altitude range of 150–250 km. The interaction of the pump wave with the ionosphere may occur strongly at the reflection altitude where the pump frequency is near the local electron plasma frequency $\omega_0 \approx \omega_{pe}$ or the upper hybrid (UH) layer where the pump frequency is near the plasma upper hybrid resonance frequency $\omega_0 \approx \omega_{uh} = \sqrt{\omega_{pe}^2 + \Omega_{ce}^2}$ where $\Omega_{ce} \approx 1.4$ MHz is the electron gyrofrequency. The upper hybrid layer is roughly 5 km or so below the reflection altitude. Many heating facilities such as EISCAT are in polar regions where the geomagnetic field B_0 is nearly vertical which implies that there are very prominent and important SEEs spectral lines produced at the upper hybrid layer. As will be discussed later, this is related to the orientation of the pump wave electric field being nearly perpendicular to B_0 at the upper hybrid layer and nearly parallel to B_0 near the reflection height. Also, it is well known that there are important changes to the spectral lines when ω_0 is swept through harmonics of the electron gyrofrequency $n\Omega_{ce}$ where for the observations discussed in detail in Section 2, $n = 2, 3$ [3].

Since 2010, a host of newly discovered phenomena associated with SEEs have been observed and studied at the High Frequency Active Auroral Research Program (HAARP) facility in Gakona, Alaska [4, 5] which is the focus of this chapter. HAARP is arguably the most powerful and versatile ionospheric heating facility presently. Its higher effective radiated power (ERP), more agile frequency and antenna array beam sweeping capabilities have opened new doors for experimental investigation of SEEs. Most of the investigations conducted at HAARP since 2010 have concentrated on SEEs within ± 1 kHz of the pump wave frequency ω_0, and often this radiation is observed within 100 Hz and even 10 Hz of ω_0. Several authors have termed the SEEs within ± 100 kHz of ω_0 wideband SEEs (WSEEs) and the more recent observations within ± 1 kHz of ω_0, narrowband SEEs (NSEEs). This later terminology will be used in the discussions here. Recent work at HAARP has also focused on the generation of SEEs near twice the pump frequency, that is, $2\omega_0$, a process known as second harmonic generation (SHG). The motivation behind these SHG investigations is two-fold. The first is to leverage the longstanding and validated techniques and theory from studies of Laser Plasma Interactions (LPI) to derive diagnostics of ionospheric plasma and the second is the presentation of an avenue to study SHG in magnetized plasmas on timescales possibly not available in LPI. Since the concentration of this chapter is on narrowband SEEs near ω_0 and $2\omega_0$, the former is referred to as PW-NSEEs and the latter SH-NSEEs, for clarity of distinction. "PW" and "SH" are respectively abbreviations for pump wave and second harmonic [6].

The nonlinear physical processes believed to initiate the generation of SEEs are parametric instabilities involving coupling of the pump electromagnetic wave into ionospheric plasma waves that can be either electromagnetic or electrostatic [7]. One of the simplest forms of a parametric instability relevant to SEEs is the three-wave parametric decay in which the pump wave decays into two downshifted sideband waves. These three waves are related by frequency and wavenumber matching conditions in which the pump wave frequency and wavenumber $(\omega_0,\ k_0)$ is equal to the sum of the frequency and wavenumber of the high frequency $(\omega_1,\ k_1)$ and low frequency $(\omega_2,\ k_2)$ decay sidebands (often an electrostatic wave), i.e.

$$\omega_0 = \omega_1 + \omega_2 \tag{1}$$

$$k_0 = k_1 + k_2 \tag{2}$$

The frequency shift between the pump frequency and the downshifted high frequency sideband, $\Delta\omega = \omega_0 - \omega_1 = \omega_2$, can be shown, through parametric instability theory, to depend on properties of the heated ionospheric plasma and therefore form the basis for diagnostic information. The temporal evolution of the high frequency sideband can also be used for diagnostic information. This evolution typically results from more strongly nonlinear processes, and comparison with nonlinear plasma simulation models can provide further possibilities for diagnostic information.

It is particularly important that the electric field vector of the pump wave on the ray path is nearly perpendicular to the geomagnetic field at the UH layer and nearly parallel near the reflection altitude for the experiments to be discussed. The pump electric field vector direction determines which wave modes participate in the parametric instability. For instance, at the UH layer, ionospheric plasma wave modes that propagate perpendicular to the geomagnetic field are expected to be involved in the parametric decay. High frequency decay modes in this case may be upper hybrid (UH) and electron Bernstein (EB) modes and low frequency wave modes may include lower hybrid (LH), ion Bernstein (IB), and highly oblique ion acoustic (IA). Interactions at the reflection altitude may include high frequency Langmuir waves (LW), a.k.a. electron plasma waves, and low frequency ion acoustic (IA) and oblique electrostatic ion cyclotron (EIC) waves since all these waves propagate with significant wave vector component along B_0. Knowledge of these wave modes will be important for interpreting the observations as well as utilizing the spectral line frequency shifts for diagnostic purposes using parametric instability theory. Since there are other nonlinear processes associated with the parametric instability such as strong turbulence development and electron heating, computational plasma modeling can be important for interpreting temporal evolution of the frequency sidebands and their association with various nonlinear ionospheric irregularity development. Therefore, the two tools, theory and modeling are critical for development of NSEE into a viable ionospheric diagnostic.

The organization of this chapter is as follows. First, recent observations of NSEE spectral lines at the HAARP facility will be discussed as well as some ancillary diagnostics. Next theoretical plasma physics concepts to explain the new SEE spectral lines will be discussed. This will largely be consideration of parametric decay instability theory. Growth rates and frequency shifts $\Delta\omega$ will be discussed and their dependence on ionospheric parameters that may provide diagnostic information. This will be followed by a discussion of nonlinear plasma simulations to provide deeper insight for linking specific NSEE spectral lines to nonlinear processes in the ionosphere such as electron acceleration and strong turbulence development. Finally, a summary and future outlook and opportunities will be provided.

2. Observations

All the observations discussed in this section were made using transmissions from the High Frequency Active Auroral Research Program (HAARP) facility in Gakona, Alaska, with location site (62.39° N, 145.15° W). The observations here were reported during the period roughly between 2010 and 2021. The maximum HAARP ERP used was 1 GW with the pump power variation depending on the experiment and O-mode polarization was used. The HAARP transmitter has approximately a 30 dB gain. The SEE signals were typically received with a 30 m folded dipole antenna located roughly 10 km from the transmitter site at the Riverview Inn. The antennas were connected to high sensitivity wideband software defined radio reviewers. The receiver shifts the frequency of the acquired signal by the transmit (pump) frequency ω_0 by mixing, and sampling it at 250 kHz. The acquired in-phase and quadrature-phase (I and Q) data are post-processed by utilizing Blackman windowing and then fast Fourier transformed (FFT) to obtain spectrograms of the received signal. The frequency resolution of the spectra was in the range of approximately 1–4 Hz depending on the experiment. The magnetic zenith direction is 14^0, that is, near vertical, with the HAARP antenna beam pointed at or near magnetic zenith with azimuth near 200^0 which was shown to provide good NSEE observations for most experiments. The HAARP transmitter frequency ω_0 was typically frequency stepped within approximately ± 100 kHz of $2\Omega_{ce} \approx$ 2.8 MHz and $3\Omega_{ce} \approx$ 4.2 MHz to provide strong interactions at the upper hybrid layer as well as interaction at the reflection altitude. The receivers measured the NSEE signals near ω_0 and also near 2 ω_0 for considering second harmonic generation. **Figure 1** shows a diagram of a typical experimental set up.

2.1 Stimulated Brillouin scatter

The first key discovery in NSEE was Stimulated Brillouin Scatter (SBS). The first observations were reported by Norin et al. [8] and soon after by [9, 10]. The latter work provided more detailed theoretical analysis that was in relatively good agreement with the behavior during the experiments. Although SBS had been studied relatively extensively within the Laser Plasma Interaction (LPI) community for several decades, a key difference was the importance of the geomagnetic field B_0 during the ionospheric heating experiments which produced new spectral lines. Also since there were pump wave interactions at both the upper hybrid and reflection altitudes, there were spectral lines believed to be associated with both interaction altitudes. The primary spectral lines observed during ionospheric heating experiments are associated with ion acoustic (IA) and electrostatic ion cyclotron (EIC) low frequency decay wave modes. The former is more prominent and produces a sideband typically shifted below ω_0 by roughly 10 Hz for interaction at the reflection altitude [10]. Sidebands shifted below the pump by several 10's of Hz can be argued to result from interaction at the upper hybrid layer by consideration of wave matching conditions. Upshifted

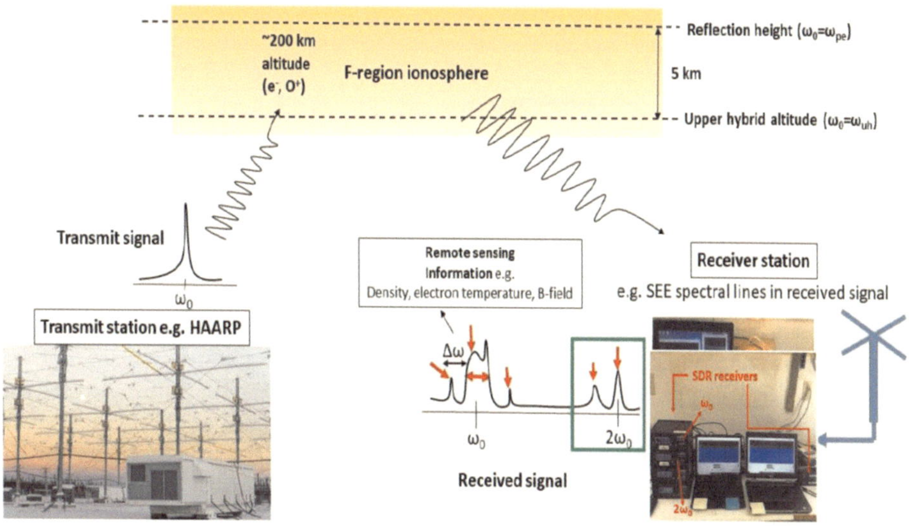

Figure 1.
A typical experimental set up to measure NSEE using transmissions from the HAARP facility. The transmitted signal produces nonlinear processes in the ionosphere between the upper hybrid layer and reflection altitude. These nonlinear processes produce secondary radiation (NSEE) observed on the ground with frequency sidebands near ω_0 and $2\omega_0$ that may be used for diagnostic information on the heated ionosphere.

sidebands could also be accounted for by wave mixing of the upwards and downwards going waves. Since the shift of the IA sideband below the pump is related to the ion acoustic frequency, which depends on temperature, it was proposed that SBS could be used as an important temperature diagnostic and there has been recent progress in this area by using comparisons with temperature measurements using Incoherent Scatter Radar (ISR) [11]. The SBS IA spectral line has also been replicated at the European Incoherent Scatter Radar (EISCAT) facility near Tromso Norway [12]. Another important feature of the IA decay line is that it can often be comparable or even larger than the pump spectral line as was noticed in the early reports. Nonlinear computer simulations show that this is consistent with the behavior of nonlinear development of the SBS parametric decay instability. Several related characteristics and phenomena are discussed in Section 4.1.

The second important SBS line was discovered by Bernhardt et al. [9] guided by the early theoretical calculations. The EIC spectral line shift is near the O^+ cyclotron (or gyro) frequency $\Omega_{cO} \approx 50$ Hz. Note O^+ is the dominant ion species in the interaction region. In general, obliquely propagating EIC waves only exist in a magnetized plasma at frequencies slightly above the ion gyrofrequency Ω_{ci} and therefore the presence of this spectral line has resulted in SBS in the ionospheric plasma being referred to Magnetized SBS (MSBS). Since in general, Ω_{ci} depends on the ion mass, i.e., $\Omega_{ci} = eB/m_i$ where e and m_i are the unit electron charge and ion mass, the importance of this spectral line to ionospheric mass spectrometry measurements was recognized. Further details will be discussed in Section 2.2.

The EIC spectral line was observed to be present when the HAARP transmit antenna beam angle was at least a few degrees off the magnetic zenith angle of 14^0 which allows the pump electric field vector to couple more effectively to the obliquely propagating EIC waves. This will be discussed in more detail in Section 3.1. Also, the EIC line requires a much higher pump power to excite. Typically, this spectral line is observed only when the pump power is above several 100 MW ERP which is considerably more than the IA spectral line which may be observed for a pump power of less than 100 MW ERP.

Figure 2 shows observations of SBS proposed to be produced near the reflection height by using ray tracing calculations [13]. The left panel shows variation of the HAARP beam angle from magnetic zenith to 7° and 14° off magnetic zenith using 1 GW ERP. The IA spectral line can be seen to be shifted by roughly 10 Hz below the pump frequency of 4.1 MHz ($\approx 3\Omega_{ce}$) and is consistent with generation at the reflection altitude. As the beam angle is increased off magnetic zenith by 14°, the EIC spectral line can be seen to appear albeit at a much lower amplitude than the IA line. The frequency shift is near $\Omega_{cO} \approx 50$ Hz. An upshifted IA spectral line can also be observed and is explained due to mixing of upwards and downward propagating wave modes [13]. The right panel shows SBS observations with varying pump power from 520 through 800 MW, to 1 GW with the HAARP antenna beam angle being 14° off magnetic zenith. It can be seen that considerably more pump power is necessary to excite the EIC spectral line, relative to the IA line, which begins to be observed above 800 MW power for these observations.

Increasing the pump power of course produces more strongly nonlinear processes in the ionosphere. Therefore, it is helpful to consider the impact of increasing pump power on the NSEE spectral line characteristics. Variation i n spectral lines between low and high pump power may potentially provide useful diagnostic information. A study of the IA spectral line characteristic variation with pump power has been made by Yellu et al. [14]. **Figure 3** shows NSEE spectra for a relatively low ($\sim$150 MW) transmit power and high ($\sim$280 MW) transmit power. The antenna beam angle was along magnetic zenith and $\omega_0 \approx 3\Omega_{ce} \approx$ 4.32 MHz. The lower power case shows a relatively distinct SBS IA line shifted below ω_0 by roughly 9 Hz and the spectral bandwidth is roughly 5 Hz. This is consistent with the observations shown earlier in **Figure 2**. Along with a considerably larger sideband spectral power, the higher power case shows a broadened line which, upon careful examination, can be described as a cascade of individual SBS IA lines. The bandwidth in the high-power case is roughly 14 Hz. Yellu et al. [14] suggests that this behavior for high power results from development of strong ion acoustic turbulence in which the backscattered EM wave

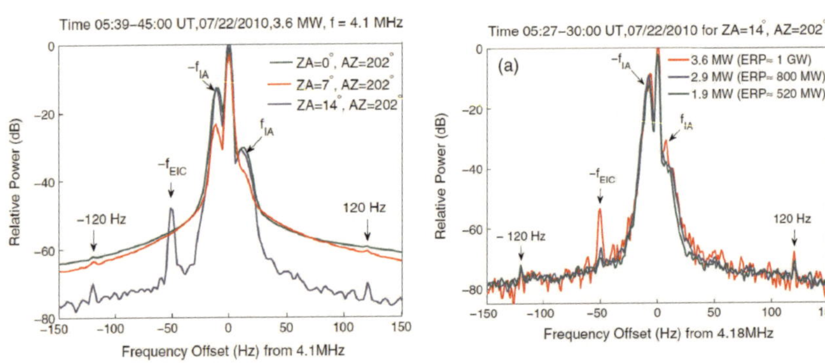

Figure 2.
Frequency spectra of stimulated Brillouin scatter (SBS) observed during ionospheric heating experiments. Two primary spectral lines exist, the ion acoustic (IA) and the electrostatic ion cyclotron (EIC). The EIC has a higher power threshold and is present for high pump powers and/or off magnetic zenith pumping. Adapted from [13].

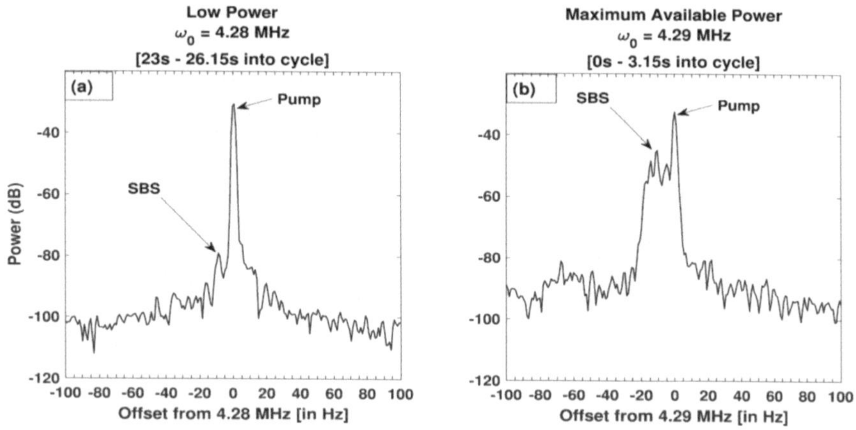

Figure 3.
Variation of IA SBS spectral line with pump power. Lower 150 MW transmit power (left) and higher 280 MW transmit power (right). Higher power produces a broader bandwidth spectral line which is expected to be linked to strong IA turbulence. Adapted from [14].

decay product of an SBS instability acts as a pump to initiate a successive SBS instability. Such strongly turbulent processes can be studied in detail using nonlinear plasma simulations as will be discussed in Section 4.1.

2.2 Ion gyroharmonic spectral lines

As was discussed in the previous section, a prominent SBS line is the EIC line which is shifted from ω_0 by the O^+ gyrofrequency Ω_{cO}. As the pump frequency is swept very near a multiple of the electron gyrofrequency $\omega_0 \approx n\Omega_{ce}$ another class of NSEE spectral lines shifted by multiples of Ω_{cO}, that is, $\Delta\omega \approx n\Omega_{cO}$, are observed. These were originally reported by Bernhardt et al. [15] with considerable observations, theory, and computer plasma modeling to follow [16–19]. This section will provide a brief summary of these NSEE spectral line observations. These NSEE spectral lines are proposed to be produced near the upper hybrid layer since it has been known that a strong pump interaction exists for $\omega_0 \approx n\Omega_{ce}$ which may produce ionospheric Artificial Field Aligned Irregularities (AFAIs) which are important for production of SEE spectral lines [3]. Also Fu et al. [19] has observed the transition from SBS in the previous section to these ion gyroharmonic spectral lines when ω_0 is stepped within a few 10's of kHz of $2\Omega_{ce}$. These NSEE spectral lines involve ion Bernstein (IB) low frequency decay wave modes. IB modes propagate nearly perpendicular to the magnetic field as so-called plasma flute mode waves [7]. They are harmonically ordered by Ω_{ci} in a hot plasma. The EIC wave can be considered as a specific type of IB wave that propagates considerably more obliquely to the magnetic field. These NSEE spectral lines were originally termed Stimulated Ion Bernstein Emissions (SIBE) [15] and later Stimulated Ion Bernstein Scatter (SIBS) due to the possible connection to SBS as demonstrated by Fu et al. [19]. For consistency with the later works, the term SIBS will be used here as necessary. These spectral line shifts of course depend on ion mass and therefore information is available on the ion species present in the heating interaction region as will be discussed.

2.2.1 Oxygen spectral lines

Since the dominant ion species at the pump interaction altitude is atomic oxygen, O^+, it is expected that the most common ion gyroharmonic spectral lines observed will be associated with O^+ and this is the case. An example of these spectral lines are observed in the left panel of **Figure 4**. These observations are at a pump power of 1 GW ERP and $\omega_0 \approx 2\Omega_{ce} \approx 2.8$ MHz. The HAARP antenna beam is pointed along the magnetic zenith 14°. It can be observed that unlike the SBS EIC line which just exhibits a single spectral line shifted below ω_0 by $\Omega_{cO} \approx 50$ Hz, there are many harmonics shifted below ordered by $n\Omega_{cO}$ with some harmonics shifted above ω_0 as well. These observations show 9 or so harmonics, however, other observations have shown 20 to 30 harmonics or more. The larger the number of harmonics has been shown to be linked to strong bulk heating of electrons across the magnetic field as will be discussed in Section 4.2. Also, the downshifted harmonics for $n > 1$ are shifted roughly near half harmonics, i.e. $(n + 1/2)\Omega_{cO}$. It has been shown that the relative amplitude of the individual spectral lines can be linked to the proximity of ω_0 to $2\Omega_{ce}$ with the lower harmonics having a larger amplitude for ω_0 very close to $n\Omega_{ce}$ in general [16] as observed in **Figure 4**.

It is observed that with increasing pump amplitude or a pump electric field vector more oblique to B_0, possibly indicating an interaction altitude further above the upper hybrid layer and closer to the reflection altitude, the line spectrum shown on the left of **Figure 4** transitions to a broadband continuous spectrum similar to the right of **Figure 4**. This transition can be replicated with theory as discussed in Section 3.2. The low frequency decay wave mode in this case is an oblique ion acoustic wave propagating 10° or so off perpendicular to B_0 near the upper hybrid layer. Such an oblique ion acoustic wave mode has linkage to the dispersion characteristics of IB wave modes and this provides insight into the nature of this transition. Also, this broad continuous spectrum can be observed on occasion with absorption lines ordered by Ω_{cO}. It can be shown that such a broad continuous spectral line indicates development of strong turbulence and electron acceleration along the magnetic field at the upper hybrid layer as will be discussed in Section 4.2.

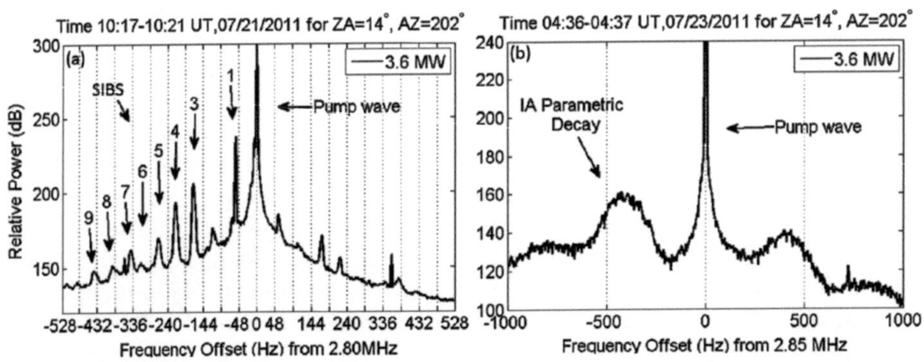

Figure 4.
HAARP observations of NSEE spectral lines shifted from ω_O by the O^+ gyrofrequency, $\Omega_{cO} \approx 50$ Hz, for $\omega_O \approx 2\Omega_{ce} \approx 2.8$ MHz (left). A typical continuous broader band spectrum (right) resulting from higher pump power and/or pump electric field vector more off perpendicular to B_o relative to pump parameters that produce the spectrum on the left. Oblique ion acoustic decay waves propagating near the upper hybrid layer are believed to be associated with this broader band spectral line. Adapted from [16].

2.2.2 Hydrogen spectral lines

Atomic hydrogen ions H^+, or protons, are expected to be a minor ion species, with density typically less than 1%, at the interaction altitude during ionospheric heating experiments. However, during geomagnetic activity there are possibilities of proton precipitation [20, 21]. In line with the possibly of utilizing the ion gyroharmonic spectral lines as a mass spectrometer, observations of NSEE spectral lines ordered by the proton gyrofrequency will be discussed. **Figure 5** (left) shows a frequency spectrum observation taken during disturbed geomagnetic conditions as described in [20]. Spectral lines both above and below $\omega_0 = 2.9$ MHz $\approx 2\Omega_{ce}$ ordered by the proton gyrofrequency, $\Omega_{cH} \approx 800$ Hz, are observed. The first downshifted spectral line shift below ω_0 has a shift that is less than Ω_{cH}. It should be noted that the frequency shifts are greater than Ω_{cH} for the higher harmonics $n > 1$. The $n = 1$ harmonic will be considered in detail to consider its diagnostic possibilities in Section 3.2. **Figure 5** (center) shows an expanded frequency view near the $n = 1$ spectral line. It can be seen that the shift is roughly 100 Hz less than Ω_{cH}, i.e., $\Delta\omega \approx$ 700 Hz. Also, the oblique ion acoustic spectral lines as observed in **Figure 4** (right) are also observed near 300 Hz shifts. It was proposed in [20, 21] that due to the low proton density, the low frequency decay wave mode associated with this spectral line is the $H^+ - O^+$ hybrid wave which has a frequency shift slightly less than Ω_{cH}, for low proton density, as will be discussed in Section 3.2. The proximity of this spectral line to Ω_{cH} can be used to determine the proton density. Information on the relative H^+ and O^+ temperature could possibly be obtained as well. The low frequency decay modes associated with the $n > 1$ proton spectral lines are proposed to be hydrogen ion Bernstein waves with frequency shift slightly larger than Ω_{cH}. These are similar to the oxygen gyroharmonic spectral lines shifted above Ω_{cO} as discussed in Section 2.2a. These proton gyroharmonic spectral lines may also provide diagnostic information. It should be noted that on occasion, however, the $n = 1$ proton spectral line shift has been observed to be larger than Ω_{cH} as shown in **Figure 5** (right). The low frequency decay mode in this case is proposed to be a hydrogen EIC wave. It is interesting to note that the SBS oxygen EIC spectral line is observed along with this proton EIC line in **Figure 5** (right). The theory for these wave modes and possibilities for diagnostics will be discussed in more detail in Section 3.2.

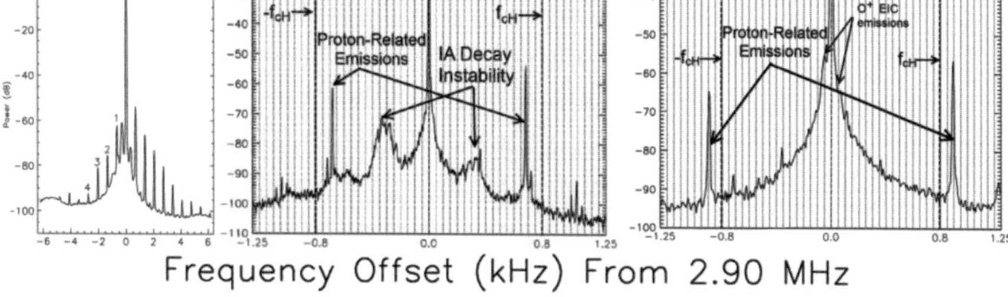

Figure 5.
NSEE spectral lines with shifts below and above the pump frequency (ω_o = 2.9 MHz) ordered by the proton gyrofrequency $\Omega_{cH} \approx$ 800 Hz (left). An expanded frequency view showing the n = 1 harmonic which has a shift below ω_o that is less than Ω_{cH}. This shift is proposed to be a diagnostic for determining the density of the minority species protons in the heated region near the upper hybrid layer (center). Another expanded view showing a spectral line shifted above Ω_{cH} believed to depend on electron and proton temperature (right). Adapted from [21].

2.2.3 Heavy metallic ion spectral lines

The previous section considered detection of light ion species, that is, protons, using NSEE spectra that exhibit spectral line shifts ordered by the ion gyrofrequency near ω_0. This section will consider the alternative of heavy metallic ions that can possibly be detected with this mass spectrometry capability. These observations were made during ionosonde detection of a sporadic E layer between 100 and 130 km during a HAARP heating campaign detailed in [22]. Although these observations are for SBS EIC lines as described in Section 2.1, there are advantages to comparing this mass spectrometry capability for heavy (relative to O^+) ions alongside that of the light ion detection in the previous section. **Figure 6** (left) shows a time-frequency spectrogram during heating where ω_0 is swept near $3\Omega_{ce}$. The frequencies shown are 4.15, 4.16, 4.17, 4.19, 4.22, 4.24, 4.29, and 4.34 MHz. There is a 30 second heating cycle shown for each of these eight frequencies.

At the altitude of 120 km, which is near the center of the observed sporadic E layer, metallic ion species expected to exist along with their gyrofrequencies are sodium (N_a^+), $\Omega_{cN_a} \approx 37$ Hz, magnesium (M_g^+), $\Omega_{cM_g} \approx 34$ Hz, calcium (C_a^+), $\Omega_{cC_a} \approx 20$ Hz, and iron (F_e^+), $\Omega_{cF_e} \approx 14$ Hz. Again, this compares to the O^+ gyrofrequency at this altitude $\Omega_{cO} \approx 49$ Hz. The observations show several important effects of stepping ω_0 near $3\Omega_{ce}$. First there is a weakening of the O^+ SBS EIC spectral line when $\omega_0 \rightarrow 3\Omega_{ce} \approx$ 4.2 MHz for the frequencies shown in the range 4.17 MHz $< \omega_0 <$ 4.22 MHz. As noted earlier, similar behavior has been observed by Fu et al. [19]. As the dominant O^+ EIC line becomes weakened, this allows another spectral line to be observed shifted below ω_0 by approximately 36 Hz as pointed out in the spectrogram. **Figure 6** (right) shows a spectrum when ω_0 = 4.17 MHz. The NSEE spectral line can again be observed more clearly downshifted by approximately 36 Hz and this has been identified as a N_a^+ EIC line and of course provides a detection of sodium ions in the sporadic E layer. It was proposed by [22] that the N_a^+ EIC line was of course generated by the interaction of the pump wave with the sporadic E layer near 120 km while the O^+ EIC line was produced near the reflection altitude interaction which was predicted by ray tracing calculations to be near 270 km. Although these observations show promise for the

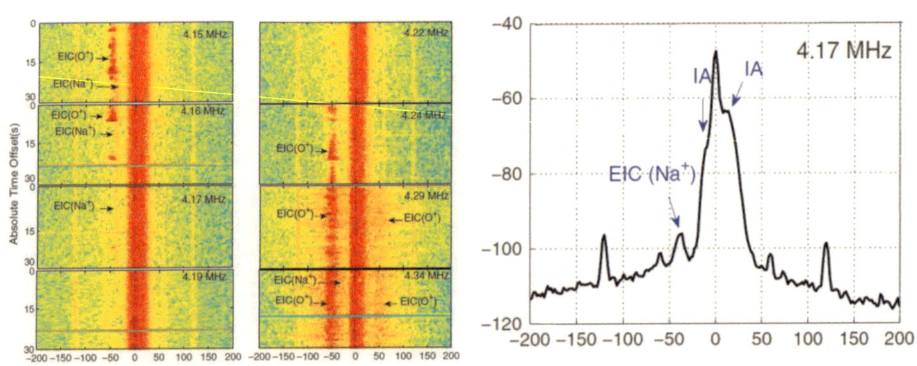

Figure 6.
Spectrogram showing SBS EIC spectral line behavior upon stepping ω_0 near $3\Omega_{ce} \approx$ 4.2 MHz (left). The corresponding power spectrum for the case when ω_0 = 4.17 MHz which reveals an EIC spectral line shifted below ω_0 by 36 Hz is shown on the right. This spectral line has been interpreted as detection of sodium ions (N_a^+) inside a sporadic E layer near 120 km altitude. Adapted from [22].

NSEE ion gyrofrequency lines to be used for detection of metallic ion species, further experiments and modeling are required. Also, the NSEE from the reflection altitude (IA and EIC spectral lines) can possibly obscure the metallic ion associated spectral lines at lower altitudes as can be seen from **Figure 6**. As was shown by the observations, stepping the pump frequency close to the $n\Omega_{ce}$ weakens the O^+ EIC line revealing the heavier metallic ion spectral lines which may be developed into a useful diagnostic. Behavior of the SHG spectral lines as ω_0 is stepped through $n\Omega_{ce}$ will be discussed in more detail in Section 2.3.

2.3 Pump second harmonic emission lines

As noted earlier, there has been recent interest in second harmonic generation (SHG) during ionospheric heating given the potential of obtaining ionospheric plasma diagnostics, for example, hydrodynamic evolution, density scale lengths, determination of interaction regions etc., as has been done in LPI, and the facility it offers to study SHG in magnetized plasmas in general. Although SHG observations had been reported in the early days of SEE research at EISCAT [3]. There were early EISCAT reports of NSEEs ordered by Ω_{cO} above and below $2\omega_0$. However, the important advantage of the SHG observations presented in this chapter is the systematic investigation of the effects of the pump wave conditioning, for example, frequency (ω_0), power, offset from $n\Omega_{ce}$ and offset of the transmit antenna beam from the magnetic zenith which is essential in investigating magnetic field effects. Also the SHG observations here are juxtaposed with NSEE within a narrowband of ω_0, that is, PW-NSEEs. This juxtaposition is important in order to properly gain insight into the wave-mixing processes underlying SHG and possibly the regions in the ionospheric plasma where the dominant interactions initiating SHG emanate from. Results from three experiments conducted at HAARP to investigate SHG are presented in this section.

The intent of the first series of experiments was to determine the power threshold for generation of SH-NSEE by varying pump power. The pump beam was pointed at a zenith angle (ZA) = 14° i.e., in the direction of the local geomagnetic field at HAARP, and an azimuth angle (AZ) of 198°. Inference from the ionogram yielded a reflection altitude of 190 km and using the International Geomagnetic Reference Field (IGRF) model yielded $3\Omega_{ce} \approx 4.324$ MHz. ω_0 was stepped near $3\Omega_{ce}$ from 4.20 to 4.36 MHz, and at each ω_0 the power was ramped from a corresponding ERP $\approx$ 28 MW to ERP $\approx$ 783 MW over a 45 s ON cycle. **Figure 7** shows PW-NSEEs (top row) and SH-NSEEs (bottom row) for progressive (left to right) sections of the pumping cycle. An SH-NSEE, labeled SH decay line, downshifted from $2\omega_0$ by $\sim 2\Delta\omega$ is seen (bottom row) where $\Delta\omega$ is the downshift of its corresponding SBS PW-NSEE line from ω_0. The PW-NSEE and SH-NSEE are observed above the noise level roughly at the same time (implying similar power thresholds) and grow progressively as the ERP increases. As previously noted in the observations of Section 2.1 and the theory that will be discussed in Section 3, $\Delta\omega$ is expected to be associated with the growth of ion acoustic waves.

The second set of experiments were designed to consider the impact of stepping ω_0 through $n\Omega_{ce}$ on SHG. Prior to providing the details of the observations, it is important to briefly note an important physical process that occurs in general as $\omega_0 \rightarrow n\Omega_{ce}$ which impacts the SEE spectrum in general. The formation of Artificial Field Aligned Irregularities (AFAIs), which are regions of deficit plasma density aligned

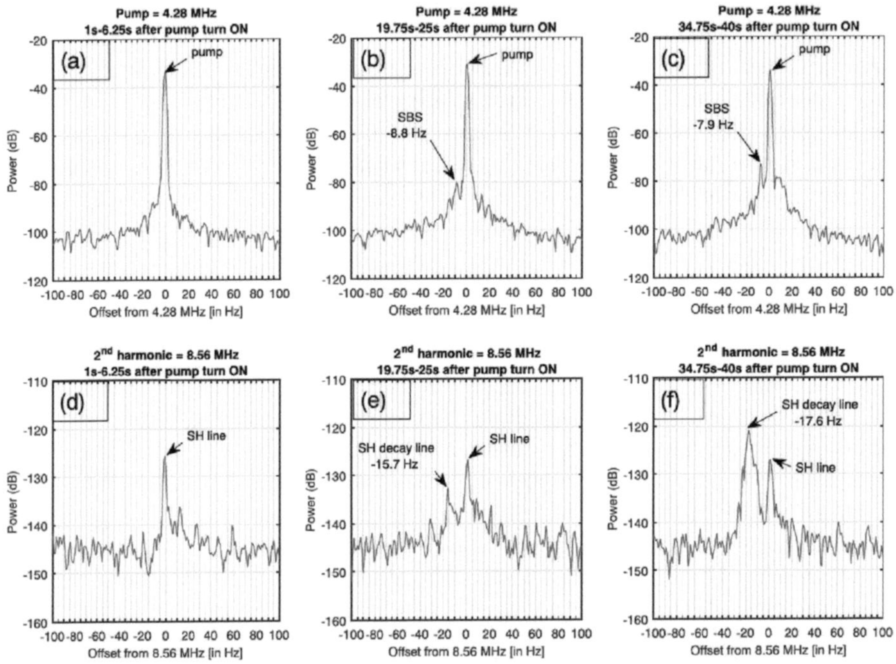

Figure 7.
SHG results showing similar power thresholds for the SH decay SH-NSEE line (lower row) to its corresponding SBS PW-NSEE line (upper row) when the pump antenna beam is along magnetic zenith. Pump ERP is ramped left to right. Adapted from [6].

along the geomagnetic field in the UH layer, produced during ionospheric heating is known to lead to a greatly attenuated pump wave or even possibly pump wave cut-off at the reflection altitude [3]. This diminishes the possibility of occurrence of SEEs in general. This was expected to likely be true for SHG proposed to be generated near the reflection altitude and was the motivation for the experiments. Therefore, the second set of experiments compared the characteristics of SH-NSEEs vis-à-vis same of PW-NSEEs for a linear power ramp from ERP $\approx$ 28 MW to maximum of ERP $\approx$ 783 MW for a 45 s pumping cycle, and for a constant maximum of ERP $\approx$ 783 for a similar duration pumping cycle. As usual, the WSEE was monitored as a proxy for AFAIs [6]. Although experimentation was done for ω_0 in the $2\Omega_{ce}$ and $3\Omega_{ce}$ regimes, the trends for the different regimes are similar and hence only results for the latter frequency regime are shown. **Figure 8** shows the suppression of the SH line and SH decay line as well as the SBS decay line upon frequency sweeping through $3\Omega_{ce} \approx 4.3$ MHz with $\omega_0 = 4.26$, 4.28, and 4.32 MHz shown. This is in line with the suppression of the SBS O^+ EIC line by frequency sweeping through $3\Omega_{ce}$ in **Figure 6**. Although not shown here, the SH-NSEE is frequency is broadened as is its corresponding PW-NSEE line for a constant high ERP pumping as shown in **Figure 3**.

The final set of observations are from an experiment that was conducted to more directly investigate the impact of the geomagnetic field, B_0, on SHG. The assessment of the impact during experimentation involved varying the pump antenna beam tilt direction relative the B_0 direction, that is, magnetic zenith (MZ), that is,

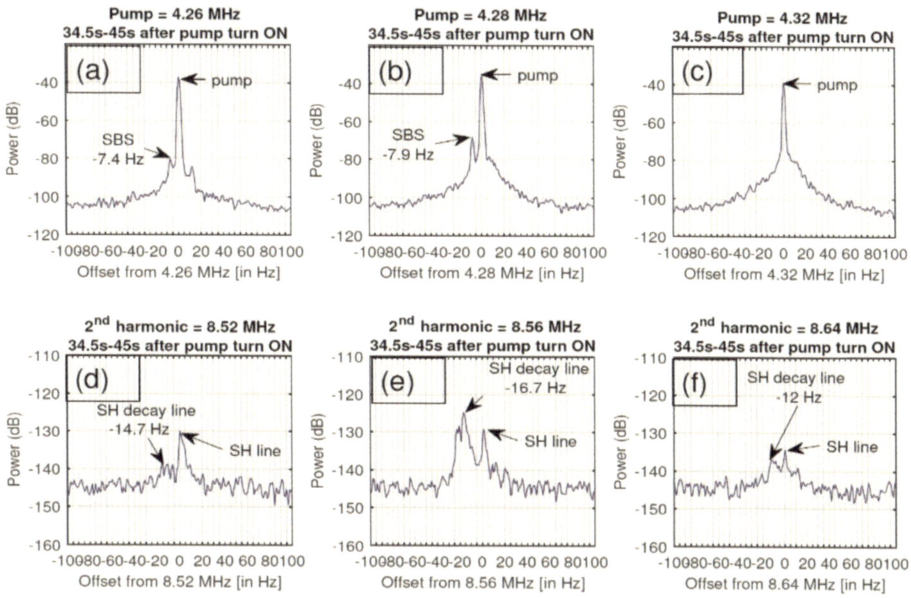

Figure 8.
SHG results showing simultaneous suppression SH decay SH-NSEE line (lower row) to its corresponding SBS PW-NSEE line (upper row) for $\omega_o \approx 3\Omega_{ce} \approx 4.324$ MHz *(rightmost column). Adapted from [6].*

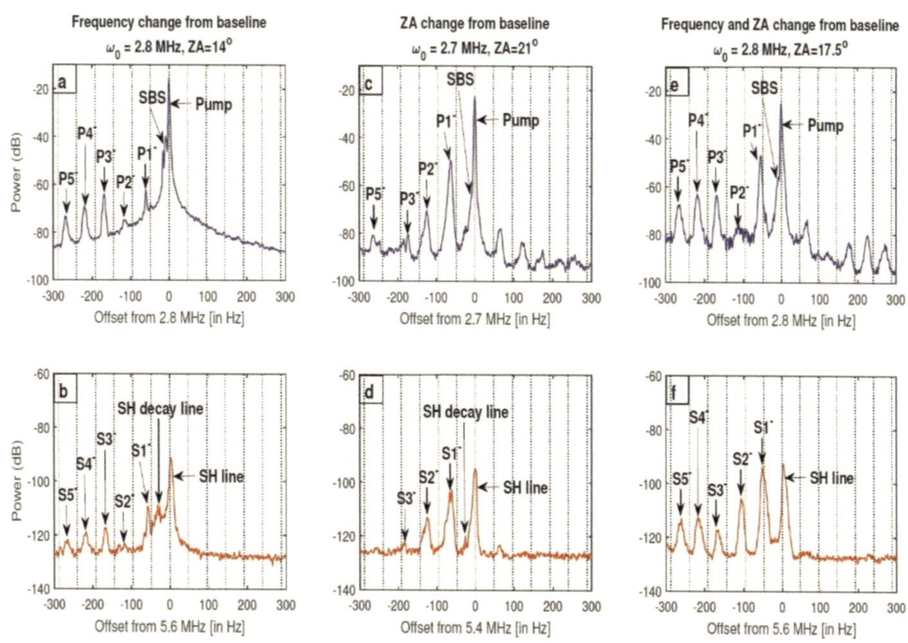

Figure 9.
SHG results showing Ω_{cO} *ordered SH-NSEEs (lower row) coincident with similarly ordered PW-NSEEs (upper row) for* ω_o *variation near* $2\Omega_{ce} \approx 2.80$ MHz *and varying pump beam tilt (ZA). Adapted from [23].*

MZ = ZA = 14°, and stepping the pump wave frequency ω_0 in the neighborhood of an electron gyrofrequency harmonic, specifically $2\Omega_{ce} \approx 2.803$ MHz. The results in **Figure 9** show Ω_{cO} ordered SH-NSEEs coincident with Ω_{cO} ordered PW-NSEEs similar to those in Section 2.2a for $\omega_0 = 2.803$ MHz $\approx 2\Omega_{ce}$ or for $\omega_0 \neq 2\Omega_{ce}$ only when pump beam is titled away from MZ, but not beyond a certain relative pump beam tilt angle [23]. In summary, these experiments indicate that proximity of ω_0 to $n\Omega_{ce}$ and antenna beam angle are two critical parameters determining the magnetic field impact on both the SH-NSEE spectrum as well as the PW-NSEE spectrum. These results also further underscore the connection of the SH-NSEE and PW-NSEE spectra that will motive and guide a qualitative theory in the next section.

3. Theory

This section will consider some basic results from plasma parametric decay instability theory to provide guidance for interpreting the observations in the previous section. This theory can provide important information on the physical processes that produce the spectral lines that are observed. Also the frequency shifts can be related to physical parameters in the ionosphere such as which species are present, density, temperature, plasma flow speed, etc. Key results such as the spectral line frequency shifts and parametric instability growth rates will be primarily summarized here with references to the details of the treatments provided. In the following discussion, it will be assumed that the frequency of the low frequency decay mode can be written as $\omega_2 = \Delta\omega + i\gamma$ where the real part of the frequency $\Delta\omega$ is equal to the frequency shift from ω_0 and γ is the growth rate of the sidebands.

3.1 Stimulated Brillouin scatter

For ionospheric Stimulated Brillouin Scatter SBS, it is assumed the pump electromagnetic wave decays into a scattered electromagnetic wave and either an ion acoustic (IA) or electrostatic ion cyclotron (EIC) wave [7]. The frequency shift of the SBS IA sideband below ω_0, as shown in **Figure 2**, can be determined by using a ray tracing model and considering the frequency and wavenumber matching conditions at the upper hybrid $\omega_0 \approx \omega_{uh}$ and resonance $\omega_0 \approx \omega_{pe}$ altitude layers for ion acoustic waves [9, 10, 13, 19]. The ion acoustic wave frequency is the frequency shift below ω_0 and depends on the angle θ (off-perpendicular to B_0) and the ion acoustic wavenumber k_{IA} determined from frequency and wave number matching on the ray path. Therefore, the shift of the IA sideband is given by

$$\Delta\omega_{IA}^2 \approx \frac{k_{IA}^2 c_{ia}^2 \cos^2\theta}{\left(1 + \frac{k_{IA}^2 c_{ia}^2}{\Omega_{cO}^2}\right)} \tag{3}$$

where $c_{ia} = \sqrt{(\gamma_e T_e + \gamma_i T_i)/m_i}$ is the ion acoustic (or sound) speed where T_e and T_i are respectively the electron and ion temperatures and $\gamma_e = 1$ and $\gamma_i = 3$. From the matching conditions $k_{IA} \approx 2k_0$, where k_0 is the pump wavenumber, and θ can be determined from the ray path. Since θ at the upper hybrid layer is relatively close to 0, the SBS ion acoustic associated sidebands generated nearer the upper hybrid layer have larger frequency shifts which is in line with interpretations of the experimental observations. Frequency shifts for the sideband generated just below the resonance altitude can be calculated from this method to be of the order of 10 Hz which is what is typically measured during experiments. The frequency shift of the EIC sideband, which is slightly above the O^+ gyrofrequency $\Omega_{cO} \approx$ 50 Hz is given by

$$\Delta\omega_{EIC}^2 \approx \Omega_{cO}^2 + \frac{k_{IA}^2 c_{ia}^2 \cos^2\theta}{\left(1 + \frac{k_{IA}^2 c_{ia}^2}{\Omega_{cO}^2}\right)} \tag{4}$$

Note the second term in Eq. (4) is assumed very small in comparison to Ω_{cO}.

It can be observed from **Figure 2** that the SBS IA spectral line has a much larger amplitude than the SBS EIC spectral line. Also the growth threshold of the SBS IA spectral line is much lower than the EIC spectral line. These characteristics can be understood at least to some degree by first considering their growth rates. Approximate analytical expressions have been provided by Shukla and Stenflo [24] and are given by

$$\gamma_{IA} \approx \frac{k_{\|} v_{osc} \omega_{pO}}{2(1 + b_O)\sqrt{\omega_0 \Delta\omega_{IA}}} \tag{5}$$

$$\gamma_{EIC} \approx \frac{k_{\perp} v_{osc} \omega_{pO}}{\sqrt{\omega_0 \Omega_{cO}}} \left[1 + \frac{b_O k_{\|}^2}{k_{\perp}^2 (1 + b_O)}\right]^{1/2} \tag{6}$$

where $b_O = k^2 c_{ia}^2 / \Omega_{cO}^2$ and ω_{pO} is the O^+ plasma frequency. $v_{osc} = eE_0/m_e\omega_0$ is the electron oscillation velocity in the pump field where E_0, m_e, e are the pump electric field strength, electron mass and unit charge. Finally, $k_{\|},\ k_{\perp}, k$ are the parallel and perpendicular components and the magnitude of the wavevector. Bernhardt et al. [10] provided a useful simplified expression for the ratio of these two growth rates $\gamma_{EIC}/\gamma_{IA} \approx \sqrt{\Delta\omega_{IA}/\Omega_{cO}} \tan\theta$ (note $\cos\theta = k_{\|}/k$). This expression implies that the IA sideband should be dominant when the pump wavevector is along magnetic zenith (i.e., along the magnetic field). However, the EIC SBS line requires that the pump wave vector be off magnetic zenith as has been observed in experiments described in Section 2.1.

Another important approach to interpreting the lower threshold of the IA sideband relative to the EIC sideband is calculation of the threshold pump electric field E_0. At this time, there has been relatively little theoretical work in this area. However, a simplified approach was presented by Mahmoudian et al. [13] which follows the concepts in [25]. The threshold field E_0 may be roughly estimated from the expression

$$\gamma_{IA,EIC} \approx \sqrt{\gamma_C \gamma_L} \tag{7}$$

where for the IA sideband $\gamma_C \approx \nu_{en}\omega_{pe}^2/2\omega_0^2$ and $\gamma_L \approx \ \Delta\omega_{IA}\sqrt{\pi/8}\sqrt{m_e/m_i}$ are the collisional and ion Landau damping rates, respectively, which are the two primary dissipative mechanisms prohibiting sideband growth. Note ν_{en} is the electron neutral collision frequency. Since the sideband growth rates are directly proportional to E_0, this threshold field can in principle be determined from Eq. (7). Mahmoudian et al. [13] considered a computation of Eq. (7) and determined a threshold field of order 1 V/m for the experiments under consideration, however, estimating the actual HAARP field amplitude in the interaction region for growth was of order 10 V/m. At this time, there are limited comparisons between observation and theory for calculation of the threshold field so this is an important future area of investigation.

The dispersion relation and growth rate expressions for IA and EIC NSEE sidebands just described in this section are valid under the cold plasma assumption. The following Section 3.2 will require plasma kinetic theory corrections to consider the dispersive characteristics of the IA and IB low frequency decay wave modes. This is necessitated by close proximity of ω_0 to $n\Omega_{ce}$ which enhances kinetic effects for interaction near the upper hybrid layer.

3.2 Ion gyroharmonic spectral lines

From Section 2.2, NSEE ion gyroharmonic spectral lines have been observed shifted from ω_0 near multiples of the O^+, H^+ and heavy metallic ion gyrofrequencies. Therefore, the parametric instability model used here is a general electrostatic model used for a multi-ion component plasma [26, 27]. The model considers a dipole (long-wavelength $k_0 = 0$) pump approximation to represent the pump electromagnetic wave. This three-wave decay model considers decay of the pump into high frequency upper hybrid/electron Bernstein modes and low frequency decay modes which are primarily ion Bernstein modes in this case. Since the model is electrostatic, the assumption is that the high frequency decay sideband modes mix with ion density fluctuations (i.e., AFAIs) to produce beat currents that ultimately result in the electromagnetic radiation observed on the ground shifted below the pump by multiples of the ion gyrofrequency [16]. Upshifted sidebands may also be produced by this mixing process.

As before, the frequency and wavenumber matching conditions are described as $\omega_0 = \omega_1 + \omega_2$; $k_0 = 0 = k_1 + k_2$ where the subscripts '0','1', and '2' denote the pump, high frequency decay mode, and low frequency ion decay mode, respectively. The parametric decay instability model of Ono et al. [27] can be stated as follows:

$$\epsilon(\omega_2) + \frac{\beta_e^2}{4}\chi_\sigma(\omega_2)\left\{\frac{\epsilon_e(\omega_2)}{\epsilon_e(-\omega_1^*)} - 2\right\} = 0 \tag{8}$$

where $(\omega_2) = 1 + \chi_e(\omega_2) + \chi_\sigma(\omega_2) = 1 + \chi_e(\omega_2) + \sum_i \chi_i(\omega_2)$, $\epsilon_e(\omega_2) = 1 + \chi_e(\omega_2)$ and $\epsilon_e(-\omega_1^*) = 1 + \chi_e(-\omega_1^*)$ with * denoting complex conjugate. Note the summation is over all the plasma ion species to provide the contribution $\chi_\sigma(\omega_s)$. The general plasma susceptibility for species j is given by

$$\chi_j(\omega) = \frac{1}{k^2\lambda_{Dj}^2}\left\{\frac{1 + \varsigma_{j0}\sum_{n=-\infty}^{n=\infty}\Gamma_n(b_j)Z(\varsigma_{jn})}{1 + \frac{i\nu_j}{k_\parallel v_{tj}}\sum_{n=-\infty}^{n=\infty}\Gamma_n(b_j)Z(\varsigma_{jn})}\right\} \tag{9}$$

where $\Gamma_n(b_j) = \mathrm{I}_n(b_j)\exp(-b_j)$, $\varsigma_{jn} = (\omega - n\Omega_j)$, $b_j = k_\perp^2 v_{tj}^2/\Omega_j^2$, $\mathrm{I}_n(b_j)$ is the modified Bessel function of first order and $Z(\varsigma_{jn})$ is the Fried Conte function. Also, $k, k_\perp, k_\parallel$ is the magnitude, perpendicular and parallel component of the wavenumber, respectively, Ω_j, v_{tj}, ν_j, and λ_{Dj} are the gyrofrequency, thermal velocity, collision frequency, and Debye length of plasma species j, respectively. The coupling coefficient proportional to the pump field E_0 is

$$\beta_e^2 = \frac{e^2}{m_e^2}\left\{\left(\frac{E_{0\parallel}k_\parallel}{\omega_0^2} + \frac{E_{0\perp}\bullet k_\perp}{\omega_0^2 - \Omega_e^2}\right)^2 + \frac{(E_{0x}k_y - E_{0y}k_x)^2\Omega_e^2}{\omega_0^2(\omega_0^2 - \Omega_e^2)^2}\right\} \tag{10}$$

where e, m_e, are the unit charge and electron mass, $E_{0\parallel}$, $E_{0\perp}$, E_{0x}, E_{0y}, k_x, k_y are the parallel, perpendicular, x and y, components of the pump electric field vector and wavenumber, respectively. Note the magnetic field E_0 is assumed in the z direction. Due to the complexity of the ion Bernstein mode structure, a Newton Raphson technique must be used to solve Eq. (8) numerically. The solution provides the real and imaginary parts of the frequency ω_2 which are equal to the frequency shift $\Delta\omega$ of the high frequency decay mode below the pump frequency ω_0 and the growth rate of this mode γ, respectively. However, an approximate analytical expression can be derived from Eq. (8) for the neutralized O^+ Bernstein harmonic wave frequency [16].

$$\Delta\omega_{nIB} \approx n\Omega_{cO}\left[1 + \frac{T_e}{T_O}\Gamma_n(b_O)\right] \tag{11}$$

Note the $n = 1$ case describes the classic cold plasma EIC wave in Eq. (4) in the long wavelength limit $b \to 0$. Also, Eq. (11) assumes propagation very close to perpendicular to B_0 ($k_\parallel/k \lesssim 0.1$). An rough expression for the wavenumber of the nth mode is $k_\perp \rho_O \sim n$ where ρ_O is the O^+ gyroradius

Figure 10 (center and right) shows calculations using Eq. (8) to characterize the transition from the discrete O^+ gyroharmonic spectral lines into a broadband oblique ion acoustic associated spectral line for increasing angle of the pump electric field vector off perpendicular to the magnetic field [16]. Again, this change in pump electric field vector orientation may be attributed to the interaction altitude being further above the upper hybrid layer and closer to the reflection altitude. The frequency shift is similar to Eq. (3) assuming the long wavelength limit and propagation near perpendicular to B_0 (i.e. $\theta \to 0$) which implies $\Delta\omega_{IA} \approx k_{IA}c_{ia}$. The growth rate γ of the sidebands is shown versus frequency shift $\Delta\omega$. The pump frequency is assumed to be near the second harmonic of the electron gyrofrequency, that is, $\omega_0 \approx 2\Omega_e$, as in the experiments of Section 2.2. The theoretical spectral line structure is in relatively good agreement with past observations as shown in **Figure 4**. The threshold electric field calculation using Eq. (8) is not shown here but has been estimated by Samimi et al. [16] to be in the range between 1 and 10 V/m. As will be noted in the following section, the broader bandwidth oblique ion acoustic spectral line can be associated with strong turbulence processes in the ionospheric upper hybrid layer including intense electron acceleration and density caviton collapse by using nonlinear plasma simulations. Therefore, observance of this NSEE spectral line may provide a very useful diagnostic for characterizing the turbulent state of the ionospheric plasma [17].

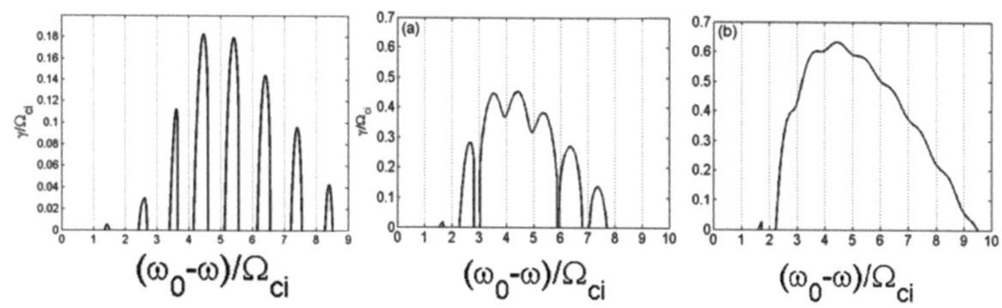

Figure 10.
*Growth rate calculations using Eq. (8) for O^+ gyroharmonic spectral lines similar to those observed in **Figure 4** (left). Note transition of the discrete line spectrum into a continuous broadband spectrum as the pump electric field vector moves further off perpendicular to the magnetic field (center and right). The continuous spectrum is in line with **Figure 4** (right). Adapted from [16].*

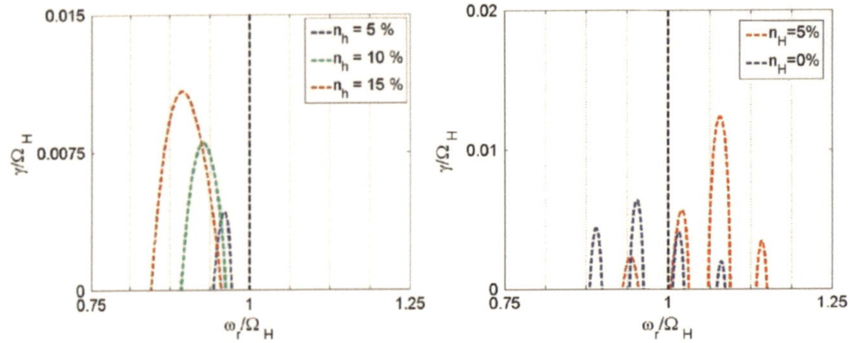

Figure 11.
*Growth rate calculations for NSEE H^+ spectral lines such as those shown in **Figure** 5 (left). Note the shift less than Ω_{cH} can be used to estimate the H^+ density from Eq. (12). Growth of H^+ spectral lines with shift larger than Ω_{cH} are also possible (right). These may provide information on H^+ temperature. Adapted from [21].*

Figure 11 (left) shows the calculations using Eq. (8) to characterize H^+ (proton) gyroharmonic spectral lines produced during ionospheric heating under the circumstances of geomagnetic activity in which proton precipitation was expected to exist [20, 21]. As noted by [20, 21] the low frequency ion decay mode in this case is the $H^+ - O^+$ hybrid wave mode that exists in the frequency band between the H^+ and O^+ gyrofrequencies. The frequency approaches the H^+ gyrofrequency as the H^+ density decreases. An approximate analytical expression (derivable from Eq. (8)) for the frequency shift of this sideband is

$$\Delta\omega_{H-O} = \Omega_{cH}(1-\Delta); \Delta \approx \left(\frac{\omega_{pH}^2}{\sum_i \omega_{pi}^2}\right)\Gamma_1(b_H)/b_H \tag{12}$$

Assuming long wavelengths ($b_H <<1$) and only two ion species, namely H^+ and O^+, then $\Delta \approx 0.5\omega_{pH}^2/\omega_{pO}^2 = 0.5(n_H m_O)/(n_O m_H)$. Therefore, the shift Δ below Ω_{cH} as observed in **Figure 5**, is directly proportional to n_H and therefore this spectral line can be used as a diagnostic for n_H. Typically, the concentration of protons is expected to be considerably less than 1% at the heating interaction altitude, however, enhancements in proton density could be detected using the frequency shift of the relatively strong H^+ NSEE spectral line shift below Ω_{cH}, i.e. $\Delta \approx 0.5(n_H m_O)/(n_O m_H) \approx$ 100 Hz [20]. Therefore, this NSEE spectral line may provide diagnostics of a minority light ion species, namely H^+ as well as an appraisal of geomagnetic activity. In the case of the observations during this experiment, the shift of the NSEE spectral line predicts a percentage of H^+ to be 10% or so. The H^+ density is also observed to be very transient (bursty) and only exists for periods of 10's of seconds at a time during the heating cycle [20, 21].

It should be noted again that H^+ spectral lines have also been observed to have frequency shifts from ω_0 slightly larger than Ω_{cH} as shown in **Figure 5** (right) [21]. This positive shift can also be described by Eq. (8) [21]. The low frequency ion decay wave is the neutralized H^+ Bernstein wave, which has a dispersive characteristic similar to Eq. (11), that has the approximate shift

$$\Delta\omega_{nIB} \approx n\Omega_{cH}\left[1 + \frac{T_e}{T_H}\Gamma_n(b_H)\right] \tag{13}$$

Also, for $n = 1$, in the long wavelength and oblique propagation regime this mode becomes a hydrogen EIC wave and implies a shift of the form similar to Eq. (4) with the substitutions $\Omega_{cO} \rightarrow \Omega_{cH}$ and $T_O \rightarrow T_H$. This wave mode is easier to drive unstable when the pump electric field vector is further off perpendicular to the magnetic field relative to the $H^+ - O^+$ hybrid wave mode responsible for spectral lines shifted less than Ω_{cH}. An example growth rate calculation for shifts larger than Ω_{cH} is shown in **Figure 11** (right). The growth rate threshold has been estimated by [21] to be between 1 V/m and 10 V/m. When there is no H^+, then only O^+ gyroharmonic associated NSEE spectral lines should be observed as in **Figure 10** (left). A small percentage of H^+, 5% in the case of the calculation, introduces H^+ gyroharmonic lines (aka hydrogen EIC for $n = 1$) with shifts larger than Ω_{cH} and amplitudes larger than the O^+ gyroharmonic spectral lines. This is consistent with observations reported by [21]. From Eq. (13), it can be seen that the shift is related to the relative H^+ temperature. Therefore, there are diagnostic possibilities for characterizing the H^+ temperature using the upshifted lines. As a final note, Eq. (8) can also be used for growth rate and threshold calculations involving the heavy ion spectral lines described in Section 2.2c. This has not been done currently and there is opportunity for future investigations.

3.3 Pump second harmonic emissions

Second harmonic generation (SHG) has been an active area of research in the Laser Plasma Interaction (LPI) community since the early 1970's [28]. Significant theoretical understanding in this community has led to the development of SHG into a useful diagnostic of the heated plasma. Using concepts from [29], the Second Harmonic (SH) line shown in **Figure 7** is taken to be generated by coalescence of the pump wave with a Langmuir wave (LW) produced by direct conversion of the pump. The mixing of these two waves of frequency ω_0 produces the observed line at $2\omega_0$. Diagnostic information can be obtained from observations of very small frequency shifts of this line from $2\omega_0$. For instance, the SH line may be used to determine the relative plasma density to the critical density ratio n_s/n_{cr} and also the bulk plasma flow speed u in the interaction region by using the small shift $\Delta\omega_{SH}$ of the SH line from $2\omega_0$, Approximate analytical expressions for this shift are given by Basov et al. [29].

$$\Delta\omega_{SH} \approx 4\frac{\omega_0 u}{c}; \Delta\omega_{SH} \approx 1 - \frac{n_s}{n_{cr}} \tag{14}$$

where c is the speed of light. Yellu et al. [14] used these expressions to estimate u and n_s in recent HAARP experiments.

SHG concepts from LPI can also be leveraged to interpret the observations shown in Section 2.3 for the Second Harmonic Decay lines (SHD). Observations were shown in **Figure 7** simultaneously for NSEE near ω_0 and NSEE near $2\omega_0$, SH-NSEE. An important observation is the SBS IA decay line which implies narrowband ion acoustic waves are present in the ionospheric plasma. An explanation of the SHD line observations can be made in line with LPI SHG, assuming for the moment that the magnetic field effects can be neglected when the pump wave antenna beam angle is along B_0 (i.e. along the magnetic zenith direction) and ω_0 is not too close to $n\Omega_{ce}$. In this case, the SHD line may be produced by a process in which the pump field decays into a LW and an IA wave. This results in a LW with frequency $\omega_0 - \Delta\omega_{IA}$. The coalescence of two of these LWs produces a sideband shifted below $2\omega_0$ by $2\Delta\omega_{IA}$ as observed in

Figure 7. It should be noted the possibility exists for coalescence of the downshifted LW with the original pump wave which would result in a SHD line with frequency $\omega_0 - \Delta\omega_{IA}$. Basov et al. [29] shows that this is possible but much less likely. This appears to be consistent with HAARP observations since the shift $\Delta\omega_{IA}$ is only observed during a small minority of experiments and these cases are under investigation. In summary, the SHD line shift can be written as

$$\Delta\omega_{SHD} \approx 2\Delta\omega_{IA} \text{ or } \Delta\omega_{SHD} \approx \Delta\omega_{IA} \tag{15}$$

with the former being the more likely to be observed.

Figure 9 also shows O^+ gyroharmic spectral lines may be produced near $2\omega_0$. It was shown for the antenna beam angle off magnetic zenith (MZ) that these lines begin to appear just as they do around the ω_0. Also, when the ω_0 is in close proximity to $n\Omega_{ce}$ these spectral lines also appear as well. An explanation of these lines near $2\omega_0$ can possibly be interpreted by leveraging the work of Tyagi et al. [30] for LPI. It is known that the O^+ Bernstein modes are produced by the three-wave parametric instability such as shown in **Figure 9**. That is, the pump wave decays into the EB/UH waves and IB waves. From [30], the pump wave with frequency and wavenumber (ω_0, k_0) exerts a nonlinear ponderomotive force of frequency and wavenumber $(2\omega_0, 2k_0)$ which in turn causes the electrons to oscillate at velocity v with frequency and wavenumber $(2\omega_0, 2k_0)$. From Eq. (11), the density fluctuations of the IB modes are described with approximate frequency and wavenumber $(\Delta\omega_{nIB}, n/\rho_O)$. These density fluctuations beat with this velocity to produce current density J with frequencies characterized by $2\omega_0 \pm \Delta\omega_{nIB}$ which produce the observed radiation on the ground. Therefore, in summary, frequency shifts for these SHD lines as observed in **Figure 9** are

$$\Delta\omega_{SHD} \approx \pm \Delta\omega_{nIB} \tag{16}$$

It should be noted that the theory presented here is very qualitative but builds on past work in LPI. Again, the magnetic field obviously plays a critical role in the SH-NSEE. There has been limited theoretical work on SHG for the ionosphere so there is much room for future investigations in theory, modeling, and also observations.

4. Modeling

Ionospheric heating with high power radio waves is an inherently nonlinear plasma process, therefore nonlinear plasma simulations are a powerful tool to interpret many of the important physical processes. The NSEE spectral lines are often observed to be highly transient during the heating cycle and exhibit increase or decrease in spectral amplitude and broadening in frequency bandwidth due to nonlinear processes. The spectral lines may also be directly related to evolution of specific nonlinear processes. The previous section considered small wave amplitude parametric decay instability theory which can be used to initially interpret observations. However, the steady state NSEE spectra will be more accurately described by nonlinear plasma simulations after saturation has occurred. The plasma simulation models discussed in this section use both the electrostatic and electromagnetic Particle-in-Cell methods with Monte Carlo collisions. These models allow the full kinetic plasma description of the nonlinear evolution of the parametric decay instabilities just discussed. Such simulation tools have long been used to investigate interaction of high power electromagnetic waves with plasmas. They will be used here to provide insight into basic nonlinear processes associated with the NSEE spectral lines.

4.1 Simulation studies of stimulated Brillouin scatter

A recent advance has been the use of computer modeling, specifically Particle-In-Cell (PIC) simulations, to investigate PW-NSEEs observed during ionospheric heating experiments. In PIC simulations, the plasma ions and electrons are modeled as simulation particles and are subject to the electric and magnetic field forces. The plasma particles, which exist within a spatially gridded domain, are moved using the field forces and weighted onto grid points. Similarly, the electromagnetic (EM) and electrostatic (ES) fields are computed using Maxwell's equations and projected to grid point values and evolve with time.

The goal of these initial simulation studies is to investigate the differences in characteristics of PW-NSEEs produced during two pump power regimes as noted in the observations described in Section 3.1. A simplified PIC simulation which approximates the essential physics for the HAARP experiments in which the observations, that is, broadband and narrowband SBS PW-NSEEs, respectively, for high and low pump power regimes, was adopted. The PIC [31] simulation model is a one space and three velocity (1D3V) dimension electromagnetic PIC simulation with Monte Carlo collisions as described in Birdsall et al. [32]. The model has plasma of uniform density in a single spatial extent in the x-direction, and is 3-dimensional for particle velocities, wave electric field and magnetic field vectors. The model is shown in **Figure 12**. Although this simulation model does not include a background magnetic field to simulate the effects of the geomagnetic field, the current model aligns with O-mode experiments at HAARP where the pump field is injected into a ZA = MZ = 14°. In this case, the nonlinear interactions that initiate PW-NSEE generation occur near the plasma resonance altitude as with the HAARP SBS PW-NSEEs observations discussed in Section 3.1. The PIC simulation model includes vacuum (moat) regions separating the plasma region from the edges of the simulation space to reduce boundary effects.

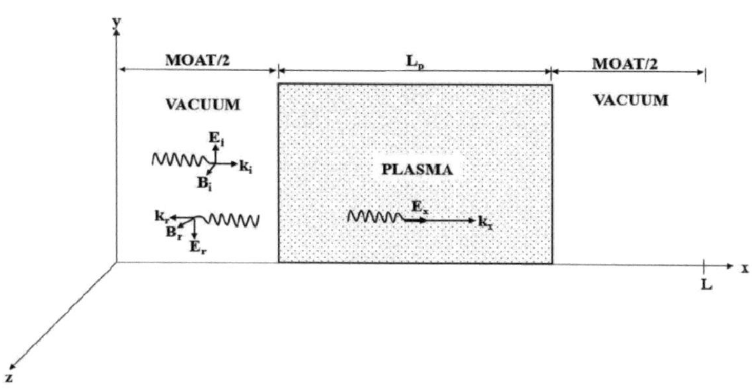

Figure 12.
PIC simulation model with a 1-D spatial extent (L) of 8192 grid cells, a plasma domain (L_p) of 4096 grid cells and left and right moat (vacuum) regions 2048 grid cells used for investigating NSEE stimulated Brillouin scatter (SBS) near the reflection altitude. (E_i, B_i, k_i) and (E_r, B_r, k_r) are the pump and reflected ELECTROMAGNETIC (EM) waves, respectively. $(E_x \| k_x)$ represents the ELECTROSTATIC (ES) low frequency ion decay waves in the plasma. Adapted from [31].

The EM pump (incident) wave (E_i, B_i, k_i) polarized in the y-direction in the adopted coordinate system is injected into the moat on the left from the left boundary. A reflected EM pump wave $(E_r B_r k_r)$ propagates back into the left moat region where its frequency spectrum is calculated and assessed for sidebands shifted from the pump frequency $\omega_0 \lesssim \omega_{pe}$. Since the simulation space as shown has aperiodic boundaries (drop-off from plasma region to plasma deficit moat regions), that is, particles that propagate outside the plasma region are not injected back, the charge densities at the boundaries are set to 0 in order to enforce periodicity and thus be able to exploit Fourier methods to compute the electrostatic (ES) field $(E_x \| k_x)$, which exists only in the plasma.

The pump electric field amplitude$|E_i|$was chosen so that the ratio of the electron oscillatory velocity in the pump electric field to the electron thermal velocity, that is, $\tilde{v}_{osc} = e|E_i|/(m_e \omega_0 v_{te})$, is reasonably consistent with a 1100 K electron temperature in the interaction region. An artifice, that is, reduced ion-electron mass ratio, commonly adopted in PIC simulations was used for the simulations to increase computational efficiency, increase resolution, but appropriately delineate the timescales of the ions and electrons involved in the wave interaction processes that produce the PW-NSEEs. **Figure 13** shows results for the low pump power case in the top two panels and the high pump power case in the bottom two panels. The ratio of the low power to high power pump electric field was taken to be 0.125. The total EM field $(E_i + E_r)$ time series is captured and averaged over a small region in the left moat. The frequency spectrum of this time series is used to be representative of the PW-NSEE observed during experiments and assess the sideband frequencies relative to the pump field. These spectra are shown in the top left and bottom left panels of **Figure 13**. The ES field time series is captured and averaged over a small region in the plasma. The spectrum of this time series is used to assess characteristics of the low frequency ion

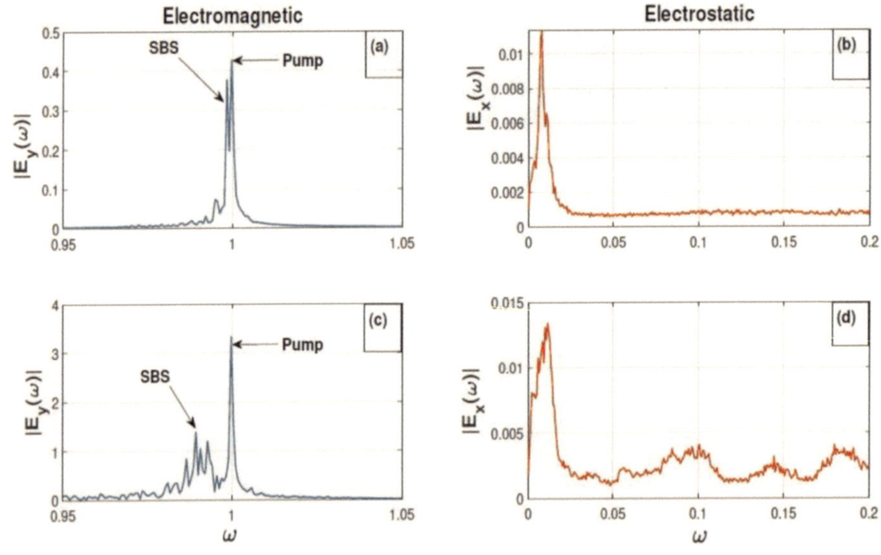

Figure 13.
PIC simulation results showing the EM field spectra (left column) and ES field spectra (right column) for low pump power (upper row) and high pump power (lower row). Frequency is normalized to ω_0. The high power case shows broadening of the spectra which is in line with the power stepping observations shown in **Figure 3**. *The ES waves are presupposed to be ion acoustic. Adapted from [31].*

decay mode. These spectra are shown in the top right and bottom right panels of **Figure 13**.

The PIC simulation results in **Figure 11** show that as with the observations from actual ionospheric heating experiments shown in **Figure 3**, the SBS PW-NSEE for the low power case (**Figure 13**a) is a distinct, narrow bandwidth spectral line whilst for high power (**Figure 13**c), it is relatively broadband. An essence of PIC simulations is the ability to probe directly the ES waves in the plasma which allows a direct means of determining the ES low frequency decay waves and therefore the specific parametric decay processes. This probe reveals corresponding narrow (**Figure 13**b) and wider bandwidth ES waves (**Figure 13**d) herein noted by virtue of proposed SBS decay causation as ion acoustic (IA) waves for the low and high-power pump regimes, respectively. The IA waves for the high-power case have a semblance of a cascade process in which initial EM waves produced in the initial parametric decay of the pump wave act as pumps to initiate subsequent parametric decays.

To investigate further the causation for the difference in PW-NSEEs spectral characteristics for the two power regimes, the ES wave (E_x) and the EM wave (E_y) in the plasma region, the ion density (n_i) and electron temperature (T_e) as determined for the high-power simulation are presented in **Figure 14**. The power in the waves, and the ion density are normalized, respectively, by the pump wave amplitude (E_0) and the ion density (n_0) at the start of the simulation. The results show that for higher power, unlike the low-power case, cavities develop in n_i in which T_e and E_y are

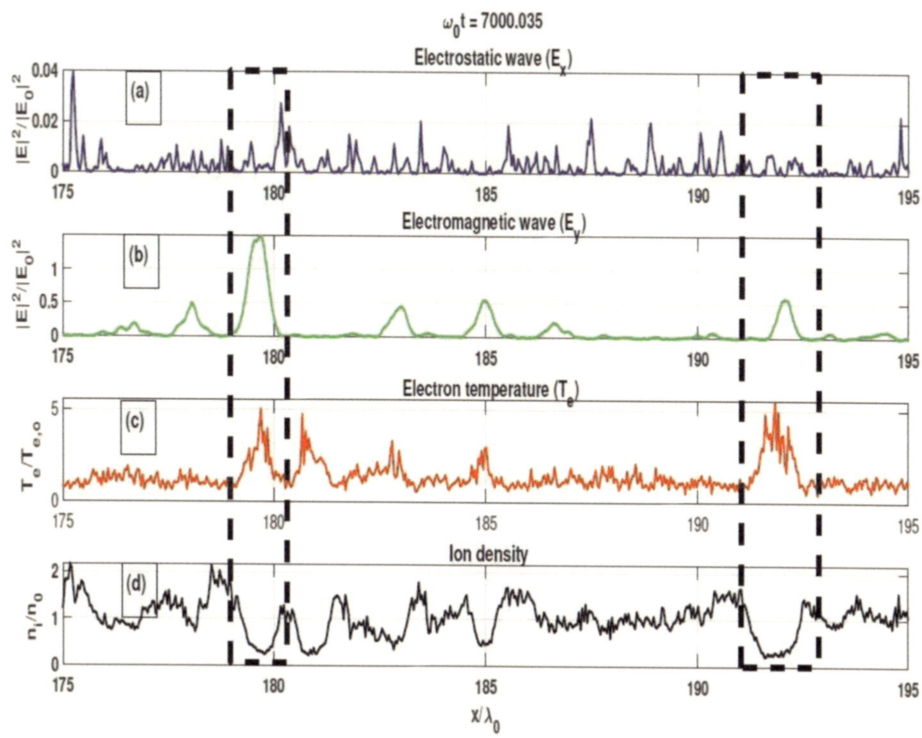

Figure 14.
ES wave (a), EM wave (b), electron temperature (c) and ion density (d) for high-power PIC simulations showing development of cavities in ion density containing high temperature electrons and enhanced EM wave amplitudes suggestive of strong ion acoustic turbulence. Adapted from [31].

significantly increased, whereas for the ES field E_x, there is not a significant correlated increase in all the cavities. Regions of particularly substantial enhancement are noted with dashed boxes in the **Figure 14**. The ES waves in the cavities are proposed to lose energy via Landau damping to heat up the plasma, thus are short-lived. The long-lived, high-power EM waves in the cavities are proposed to be indicative of strong plasma turbulence which is the causation of the broadband SBS PW-NSEEs structuring observed for high-power ionospheric heating experiments [31]. These strong EM fields trapped in the cavities imply high reflectivity and result in the amplitude of the scattered sideband having comparable or even larger amplitude to the pump as observed in heating experiments.

4.2 Simulation studies of oxygen Gyroharmonic spectral lines

Figure 15 shows the configuration of an electrostatic two-dimensional PIC model (2D3V) using Monte Carlo collisions to investigate nonlinear processes associated with NSEE O^+ gyroharmonic spectral lines near the upper hybrid layer as discussed in Section 2.2a [17]. In this case $\omega_0 \approx 2\Omega_{ce} = \omega_{uh}$. The model uses a dipole pump approximation (i.e., $k_0 \approx 0$, as does the theory in Section 3.2). Therefore, the pump electric field is applied as a uniform driver across the entire simulation domain and is of the time-varying form $E = E_0 \cos(\omega_0 t)$. The constant background magnetic field B_0 is applied in the $\hat{z}$ direction as shown. The electric field vector is varied at an angle that is off perpendicular to the magnetic field θ_E. Within this electrostatic model framework, the current density in the simulation domain would be assumed to produce the SEE observed on the ground. The first set of simulations to be discussed investigated the O^+ gyroharmonic spectral lines. Therefore, the neutralized O^+ Bernstein waves are the low frequency ion decay mode. In this case $\theta_E \gtrsim \sqrt{m_e/m_O}$. The destabilized waves propagate near the direction of the pump electric field vector. All the simulations are run to nonlinear saturation in which the field and plasma particle thermal energies have reached a nonlinear saturated state which occurs on the order of 10 growth periods.

Figure 16 (top right) shows the temporal growth of the electric field energy in the y direction, which is nearly perpendicular to B_0, for the parametric instability involving low frequency neutralized O^+ Bernstein decay waves. It can be seen that early in

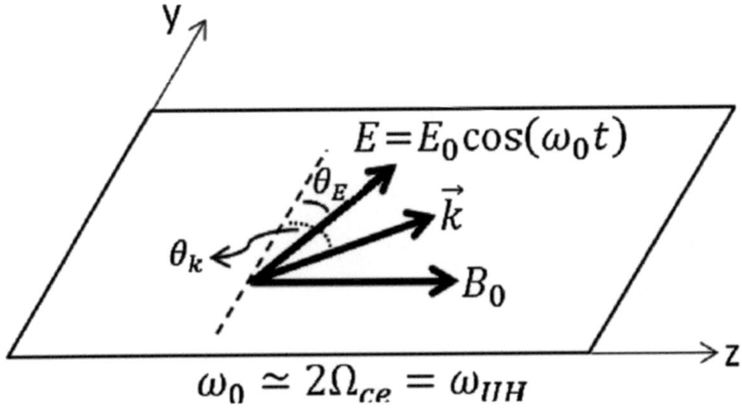

Figure 15.
Schematic diagram of 2-dimensional electrostatic PIC simulation model used to investigate generation of parametric decay instabilities that produce NSEE near the upper hybrid layer. Adapted from [17].

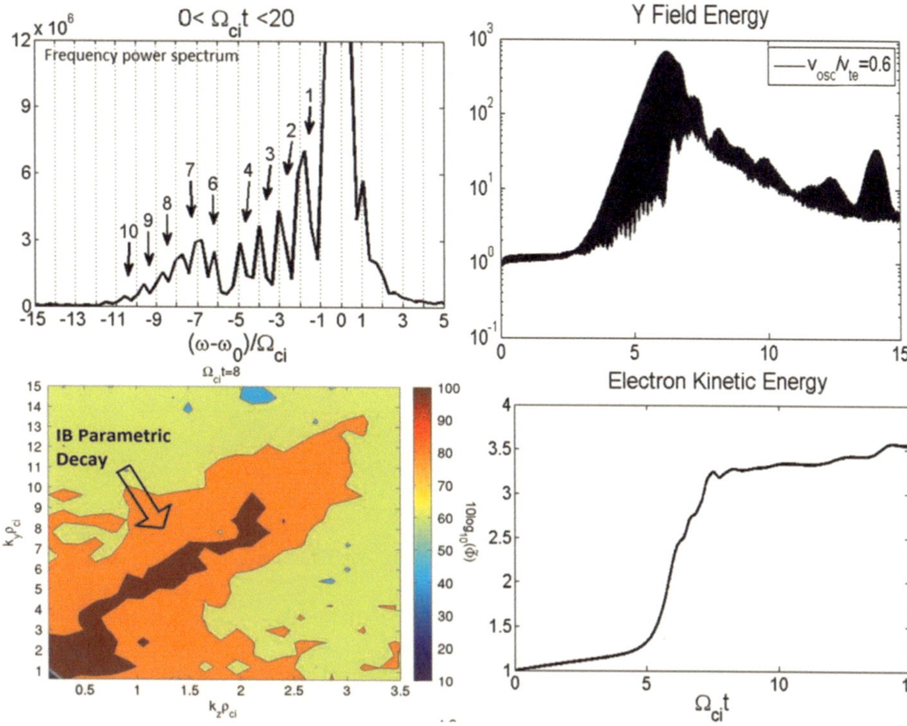

Figure 16.
Power spectrum of simulation waves showing existence of spectral lines shifted below ω_O, ordered by $n\Omega_{cO}$, after nonlinear saturation (top left) similar to ***Figure 4*** *(left). Wavenumber spectrum showing characteristic neutralized ion Bernstein modes (bottom left). Growth of electric field energy associated with the parametric instability involving neutralized ion Bernstein waves (top right). Electron kinetic energy showing bulk heating of electrons across the magnetic field associated with O^+ gyroharmonic spectral lines (bottom right). Adapted from [17].*

the simulation, the electric field energy grows linearly which is in line with small amplitude growth predicted by Eq. 8. After linear growth, the field energy eventually reaches saturation. **Figure 16** (top left) shows a frequency power spectrum of the electric field time series after the simulation reaches a saturated state. It can be seen that there are roughly 10 O^+ gyroharmonic spectral lines shifted below the pump frequency which is similar to the observation spectrum shown in **Figure 4**. The results in **Figure 16** (top left) are also in line with the theoretical predictions of Eq. (8) in **Figure 10**. A closer examination of the dispersive properties of these waves is provided in **Figure 14** (bottom left) by using a $k_y - k_z$ wavenumber spectrum. This wavenumber spectrum indicates the expected characteristic wavenumber $k_y\rho_{ci} \sim n$, n the harmonic number, which validates the production of neutralized ion Bernstein waves. It is observed in **Figure 16** (bottom right) that significant bulk heating of the electrons across the magnetic field occurs. The electron kinetic energy across the magnetic field is shown and it can be seen that this electron kinetic energy grows eventually reaching nonlinear saturation near time $\Omega_{cO}t \approx 7$ as does the electric field energy. It was observed that the number of harmonic lines observed in the simulations is related to the degree of electron heating across the field [17]. The number of harmonics, of course, increases with increasing pump power.

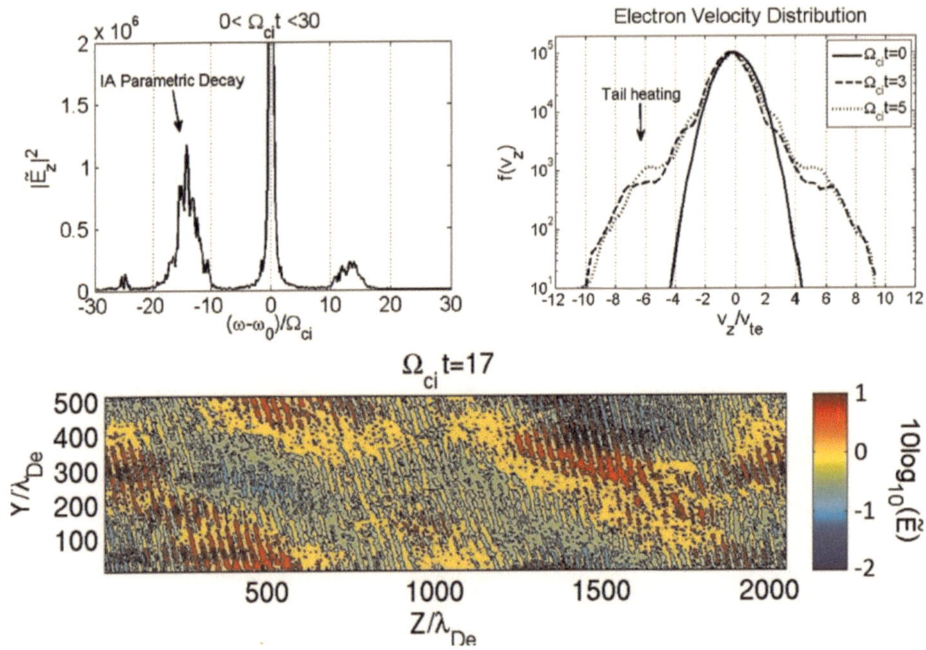

Figure 17.
Power spectrum of simulation waves showing development of a spectral line associated with ion acoustic decay waves at the upper hybrid layer (top left). This spectrum is in line with observations in ***Figure 4****. Electron acceleration and development of intense electric fields inside density cavities associated with the spectral line (top right and bottom). Adapted from [17].*

The second set of simulations performed with this model consider the generation of the broadband continuous NSEE spectral lines associated with the ion acoustic waves near the upper hybrid layer as shown in **Figure 4**. Again, rather than IB low frequency decay wave modes as in the previous simulations, the low frequency decay modes in this case are IA. This implies that the pump electric field vector must be further off perpendicular to B_0 so that $\theta_E \gg \sqrt{m_e/m_O}$. This range of θ_E allows coupling into the IA waves which propagate at a more oblique angle off perpendicular to B_0 than the IB waves that result in a discrete line spectrum. **Figure 17** shows results from simulations to consider the NSEE line shown in **Figure 4** (right) and theoretically predicted in **Figure 10** (right). **Figure 17** (top left) shows an electric field power spectrum in the simulation after nonlinear saturation. A spectral line can be seen that is downshifted from the pump frequency. The spectral line has a broad bandwidth relative to the O^+ gyroharmonic lines observed in the simulations of **Figure 14**. A sideband upshifted by approximately the same frequency can be seen as well that has a smaller amplitude. This is in line with the observations of **Figure 4**. **Figure 15** (top right) shows the evolution of the electron velocity distribution along B_0 during the simulation. Strong tail heating of the electron velocity distribution occurs. **Figure 17** bottom shows the electric field on the spatial grid during the simulation. The red areas show intense electric fields that are associated with cavities in the density similar to **Figure 12**. These intense electric fields in the cavities are responsible for the acceleration of the electrons along the magnetic field. The density structures (not shown) show caviton collapse behavior in which the cavities (and electric fields within the cavities) intensify and then collapse, transferring energy to the electrons through acceleration along the magnetic field. This is similar to elevation of the electron temperature in the cavities shown in **Figure 14**. Therefore the observation of this NSEE spectral line is a diagnostic for intense acceleration of the electrons along B_0 and development of cavities containing intense electric fields in the upper hybrid layer.

5. Conclusions

Significant progress has been made during the past decade in beginning to develop pump wave NSEE (PW-NSEE) into a useful diagnostic for the ionosphere during heating experiments. Of particular note is the possibility for investigating various plasma species during the heating experiments in terms of species mass, density, and temperature although further work must be done for development into a more robust technique in comparison to other diagnostics such as Incoherent Scatter Radar (ISR), e.g. [11]. Even with the current state of knowledge that has been surveyed here, PW-NSEE is clearly an important diagnostic that should be utilized with other mainstay diagnostics during heating experiments. At this time, investigation of heavy ions at lower altitudes, that is, Section 2.2.3, is a particularly attractive possibility due to the difficulty of making such measurements, however, considerably more observations are necessary. Also, at this time, little if any theoretical and modeling development has been performed to quantify the parametric instabilities that are at play for heavy ions, however, the framework discussed in Sections 3.1 and 3.2 could be leveraged as a starting point. Possible assessment of geomagnetic activity with PW-NSEE as noted in Section 2.2.2 requires further observational investigation as well, however, a useful theoretical framework has been provided in Section 3.1 and 3.2 that imply a useful diagnostic technique.

Although there has not been elaboration on AFAIs given the scope of this chapter, AFAIs play an important role in the generation, evolution and possibly the quenching or attenuation of PW-NSEEs and SH-NSEEs. Past experimental observations have shown enhancement of AFAIs for $\omega_0 \approx 2\Omega_e$ heating and suppression of AFAIs for $\omega_0 \approx n\Omega_e$, for n > 2, e.g. [33]. For PW-NSEEs which are proposed to originate from a parametric decay in which the high frequency decay mode is electrostatic or for SHG processes proposed to be due to the mixing of initially generated electrostatic modes, the mode or “mixed” mode must scatter from gradients in AFAIs into an EM wave in order to be observable on the ground as PW-NSEEs or SH-NSEEs, respectively. AFAIs are known to be formed near the UH altitude, and thus may inhibit SEEs proposed to be generated at the reflection altitude. The foregoing can thus be used to design heating experiments, for example, pumping $\omega_0 \approx n\Omega_e$ in different gyroharmonic regimes with high resolution frequency stepping, and observing the development and evolution of the SEE spectral lines and AFAIs. This may enable the distinction of generation altitudes and, also possibly yield confirmation of wave decay modes involved in the underlying nonlinear interaction processes.

Use of PW-NSEE as a diagnostic for nonlinear evolution and strong turbulence during heating experiments has been greatly aided by utilization of plasma simulation models, however, considerable work remains in this area as well. Spectral line characteristics and temporal evolution have been shown to be linked to specific strongly nonlinear signatures such as density cavity development, density cavity collapse, electron heating, electron velocity distribution tail acceleration, and trapping of both EM and ES waves in density cavities. These phenomena all provide insight into the heating process and nature of the interaction between the high power radiowave and ionospheric plasma. Although progress has been made, future modeling investigations are necessary. Of particular importance is the extension of the model described in Section 4.1 to include magnetic field effects which would enable more careful study of the SBS EIC spectral line. This line has been shown through observations to be particularly important as a mass spectrometer, however, observations also show important temporal evolution that are likely of great importance, e.g. [17]. At this time specific nonlinear physics associated with this spectral line are not well known so there are significant opportunities for contributions in this area. Finally, the change in PW-NSEE spectral characteristics and associated physics upon stepping ω_0 through $n\Omega_{ce}$ have been studied through observations, for example, Sections 2.2.3, 2.3, but more careful modeling work is required.

The last critical area that would benefit from considerable more investigation, observations, theory, and modeling, is Second Harmonic Generation SHG during ionospheric heating experiments, described in this chapter as Second Harmonic NSEE (SH-NSEE). Although a number of successful experiments producing useful observations have been performed over the past 5 years or so, little work has been done on theory and modeling. It is clear from the LPI community that SHG is capable of providing important diagnostic information and the same should be true for ionospheric heating. Only a qualitative theoretical model has been provided in this chapter that is based on concepts from LPIs. An extension of the model of Section 4.1 would be a plausible starting point.

In summary, the outlook for continued development of NSEE into a useful ionospheric heating diagnostic is promising. A coordinated approach of observations, theory, and modeling is critical to exploit the full potential of this interesting new area of investigation in ionospheric research.

Acronyms and abbreviations

SEEs	Stimulated Electromagnetic Emissions
NSEEs	Narrowband Stimulated Electromagnetic Emissions
PW-NSEEs	Pump Wave NSEEs
SH-NSEEs	Second Harmonic NSEEs
SHG	Second Harmonic Generation
HAARP	High Frequency Active Auroral Research Program
EISCAT	European Incoherent Scatter Radar
SBS	Stimulated Brillouin Scatter
MSBS	Magnetized Stimulated Brillouin Scatter
IA	Ion Acoustic
EIC	Electrostatic Ion Cyclotron
ES	Electrostatic
LH	Lower Hybrid
UH	Upper Hybrid
LW	Langmuir Wave
IB	Ion Bernstein
EB	Electron Bernstein

Author details

Wayne Scales[1*] and Augustine Yellu[2]

1 Bradley Department of Electrical and Computer Engineering, Virginia Tech, Blacksburg, Virginia, USA

2 Avionics Engineering Center, Ohio University, Athens, Ohio, USA

*Address all correspondence to: wscales@vt.edu

References

[1] Gurevich A. Nonlinear effects in the ionosphere. Physics – Uspekhi. 2007; **50**(11):1091-1121. DOI: 10.3367/UFNr.0177.200711a.1145

[2] Thide B, Kopka H, Stubbe P. Observations of stimulated scattering of a strong high frequency radio wave in the ionosphere. Physical Review Letters. 1982;**49**:1561-1564. DOI: 10.1103/PhysRevLett.49.1561

[3] Leyser T. Stimulated electromagnetic emissions by high-frequency electromagnetic pumping of the ionospheric plasma. Space Science Reviews. 2001;**98**:223-328. DOI: 10.1023/A:1013875603938

[4] Scales W. Recent advances in stimulated radiation studies during radiowave heating the near earth space environment. Radiation Effects and Defects in Solids. 2016;**171**(1–2): 2-12. DOI: 10.1080/10420150.2016.1155586

[5] Scales W. Recent advances in ionospheric stimulated electromagnetic emission investigations. In: 2018 12th International Symposium on Antennas, Propagation and EM Theory (ISAPE). Hangzhou, China: IEEE; 2018. pp. 1-4

[6] Yellu A, Scales W, Mahmoudian A, Bernhardt P, Siefring C, McCarrick M. First observations of narrowband stimulated electromagnetic emissions at the pump frequency second harmonic during ionosphere interaction experiments. Geophysical Research Letters. 2018;**45**:8690-8697. DOI: 10.1029/2018GL078924

[7] Chen F. Introduction to plasma physics and controlled fusion. In: Plasma Physics. 2nd ed. Vol. 1. New York: Springer; 1984

[8] Norin L, Leyser T, Nordblad E, Thide B, McCarrick M. Unprecedentedly strong and narrow electromagnetic emissions stimulated by high-frequency radio waves in the ionosphere. Physical Review Letters. 2009;**102**:065003. DOI: 10.1103/PhysRevLett102.065003

[9] Bernhardt P, Selcher C, Lehmberg R, Rodriguez P, Thomason J, Groves K, et al. Stimulated Brillouin scatter in a magnetized ionospheric plasma. Physical Review Letters. 2010;**104**(16):165004. DOI: 10.1103/PhysRevLett.104.165004

[10] Bernhardt P, Selcher C, Lehmberg R, Rodriguez P, Thomason J, McCarrick M, et al. Determination of the electron temperature in the modified ionosphere over HAARP using the HF pumped stimulated Brillouin scatter (SBS) emission lines. Annales de Geophysique. 2009;**27**:4409-4427. DOI: 10.5194/angeo-27-4409-2009

[11] Fu H, Jiang M, Vierinen J, Haggstrom I, Rietveld M, Varberg E, et al. A stimulated emission diagnostic technique for the high power radio wave modified ionosphere. Geophysical Research Letters. 2022;**49**: e2022GL099390. DOI: 10.1029/2022GL099390

[12] Fu H, Scales W, Bernhardt P, Briczinski S, Kosch M, Senior A, et al. Stimulated Brillouin scattering during electron gyro-harmonic heating at EISCAT. Annales Geophysicae. 2015;**33**: 983-990. DOI: 10.5194/angeo-33-983-2015

[13] Mahmoudian A, Scales W, Bernhardt P, Fu H, Briczinski S, McCarrick M. Investigation of ionospheric stimulated Brillouin scatter generated at pump frequencies near electron gyroharmonics. Radio Science.

2013;**48**:685-697. DOI: 10.1002/2013RS005189

[14] Yellu A, Scales W, Mahmoudian A, Bernhardt P, Siefring C, Briczinski S, et al. Pump power effects on second harmonic stimulated electromagnetic emissions during ionosphere heating. Journal of Geophysical Research: Space Physics. 2019;**124**:9739-9754. DOI: 10.1029/2019JA027282

[15] Bernhardt P, Selcher C, Kowtha S. Electron and ion Bernstein waves excited in the ionosphere by high power EM waves at the second harmonic of the electron cyclotron frequency. Geophysical Research Letters. 2013;**38** (L19107). DOI: 10.1029/2011GL049390

[16] Samimi A, Scales W, Fu H, Bernhardt P, Briczinski S, McCarrick M. Ion gyroharmonic structures in stimulated radiation during second electron gyroharmonic heating: 1. Theory. Journal of Geophysical Research: Space Physics. 2013;**118**:502-514. DOI: 10.1029/2012JA018146

[17] Samimi A, Scales W, Bernhardt P, Briczinski S, McCarrick M. Ion gyroharmonic structures in stimulated radiation during second electron gyroharmonic heating: 2. Simulations. Journal of Geophysical Research: Space Physics. 2014;**119**:462-478. DOI: 10.1002/2013JA019341

[18] Mahmoudian A, Scales W, Bernhardt P, Samimi A, Kendall E, Ruohoniemi J, et al. Ion gyro-harmonic structuring in the stimulated radiation spectrum and optical emissions during electron gyro harmonic heating. Journal of Geophysical Research: Space Physics. 2013;**118**:1270-1287. DOI: 10.1002/jgra.50167

[19] Fu H, Scales W, Bernhardt P, Samimi A, Mahmoudian A, Briczinski S, et al. Stimulated Brillouin scatter and stimulated ion Bernstein scatter during electron gyroharmonic heating experiments. Radio Science. 2013;**48**: 607-616. DOI: 10.1002/2013RS005262

[20] Bordikar M, Scales W, Samimi A, Bernhardt P, Brizcinski S, McCarrick M. First observations of minority ion (H+) structuring in stimulated radiation during second electron gyroharmonic heating experiments. Geophysical Research Letters. 2013;**40**:1479-1483. DOI: 10.1002/grl.50327

[21] Bordikar M, Scales W, Mahmoudian A, Kim H, Bernhardt P, Redmon R, et al. Impact of active geomagnetic conditions on stimulated radiation during ionospheric second electron gyroharmonic heating. Journal of Geophysical Research: Space Physics. 2014;**119**:548-565. DOI: 10.1002/2013JA019367

[22] Mahmoudian A, Scales W, Bernhardt P, Isham B, Kendall E, Briczinski S, et al. Electron gyroharmonic effects on ionospheric stimulated Brillouin scatter. Geophysical Research Letters. 2014;**41**:5710-5716. DOI: 10.1002/2014GL061050

[23] Yellu A, Scales W, Bernhardt P, Siefring C, Briczinski S, Nossa E, et al. Geomagnetic field impacts on second harmonic generation during high power radio wave-ionosphere interaction. Physics of Plasmas. 2021;**28**:062901. DOI: 10.1063/5.0031588

[24] Shukla P, Stenflo L. Stimulated Brillouin scattering of electromagnetic waves in magnetized plasmas. Journal of Plasma Physics. 2010;**76**:853-855

[25] Kruer W. The Physics of Laser Plasma Interactions. New York: Addison-Wesley; 1988

[26] Porkolab M. Theory of parametric instability near the lower-hybrid frequency. Physics of Fluids. 1974;**17**: 1432-1442. DOI: 10.1063/1.1694910

[27] Ono M, Porkolab M, Chang R. Parametric decay into ion cyclotron waves and drift waves in multi-ion species plasma. Physics of Fluids. 1980; **23**:1656. DOI: 10.1063/1.863185

[28] Caruso A, DeAngelis A, Gatti G, Gratton R, Martellucci S. Second-harmonic generation in laser produced plasmas. Physics Letters A. 1970;**33**(1): 29-30

[29] Basov N, Bychenkov V, Krokhin O, Osipov M, Rupasov A, Silin V, et al. Second harmonic generation in a laser plasma. Soviet Journal of Quantum Electronics. 1979;**9**:1081

[30] Tyagi Y, Tripathi D, Walia K. Laser second harmonic generation in a magnetoplasma assisted by an electrostatic wave. Physics of Plasmas. 2017;**24**:043104

[31] Yellu A, Frazier J, Scales W. Simulation studies of strongly turbulent stimulated Brillouin backscattering during ionospheric heating. Radiation Effects and Defects in Solids. 2020;**175** (1–2):141-149. DOI: 10.1080/10420150.2020.1718139

[32] Birdsall C, Langdon A. Plasma Physics Via Computer Simulation. New York: Institute of Physics; 1991

[33] Djuth FT, Pedersen TR, Gerken EA, Bernhardt PA, Selcher CA, Bristow WA, et al. Ionospheric modification at twice the electron cyclotron frequency. Physical Review Letters. 2005;**94**(12): 125001